T0291840

CAMBRIDGE LIBRARY COLLECTION

Books of enduring scholarly value

Technology

The focus of this series is engineering, broadly construed. It covers technological innovation from a range of periods and cultures, but centres on the technological achievements of the industrial era in the West, particularly in the nineteenth century, as understood by their contemporaries. Infrastructure is one major focus, covering the building of railways and canals, bridges and tunnels, land drainage, the laying of submarine cables, and the construction of docks and lighthouses. Other key topics include developments in industrial and manufacturing fields such as mining technology, the production of iron and steel, the use of steam power, and chemical processes such as photography and textile dyes.

Personal Narrative of the Origin and Progress of the Caoutchouc or India-Rubber Manufacture in England

Over the course of three decades, the English businessman and inventor Thomas Hancock (1786–1865) took out sixteen patents relating to the potential applications of natural rubber. Hancock's fascination with this material, the properties of which had not been fully appreciated in England, drove him to experiment extensively with manufacturing methods. This led to the development of the mechanical process of 'mastication' and the chemical process of vulcanization, the end product of which was used by Macintosh to create waterproof garments. This illustrated account of Hancock's discoveries and methods was first published in 1857. It covers the origin of his interest in natural rubber and his subsequent experiments and patents. Also included are details about the plants from which natural rubber is extracted. The result is an informed chronicle of the commercial exploitation of a versatile and lucrative resource.

Personal Narrative
of the Origin and Progress
of the Caoutchouc
or India-Rubber
Manufacture in England

THOMAS HANCOCK

CAMBRIDGE
UNIVERSITY PRESS

CAMBRIDGE
UNIVERSITY PRESS

University Printing House, Cambridge, CB2 8BS, United Kingdom

Published in the United States of America by Cambridge University Press, New York

Cambridge University Press is part of the University of Cambridge.

It furthers the University's mission by disseminating knowledge in the pursuit of
education, learning and research at the highest international levels of excellence.

www.cambridge.org
Information on this title: www.cambridge.org/9781108069281

© in this compilation Cambridge University Press 2014

This edition first published 1857
This digitally printed version 2014

ISBN 978-1-108-06928-1 Paperback

HANHART, IMP

Tho Hancock
1856

Personal Narrative

OF

THE ORIGIN AND PROGRESS

OF THE

CAOUTCHOUC

OR

INDIA-RUBBER MANUFACTURE

IN

ENGLAND.

By THOMAS HANCOCK,

OF THE FIRM OF

CHARLES MACINTOSH AND CO.
London and Manchester.

WITH ENGRAVINGS.

TO WHICH IS ADDED SOME ACCOUNT OF THE

PLANTS FROM WHICH CAOUTCHOUC IS OBTAINED,

ITS CHEMICAL ANALYSIS, STATISTICAL TABLES, ETC. ETC.

WITH

An Appendix,

CONTAINING

THE SPECIFICATIONS OF THE AUTHOR'S PATENTS.

LONDON:
LONGMAN, BROWN, GREEN, LONGMANS, & ROBERTS.
1857.

London:
Printed by Spottiswoode & Co.,
New-street-Square.

The material originally positioned here is too large for reproduction in this reissue. A PDF can be downloaded from the web address given on page iv of this book, by clicking on 'Resources Available'.

PREFACE.

In writing a *personal* narrative, it is impossible to escape the very disagreeable necessity of frequently repeating the pronoun I,—my readers must excuse this unavoidable egotism.

This humble relation is in no respect to be considered a treatise on Caoutchouc; it is simply an account of my own progress in the manufacture— feeling conscious that no one had preceded me in this path, or I should not have assumed as much in the Title-page.

This claim is not of recent date, as my Patents will show, and the notices of my manufactures from time to time in the "Mechanic's Magazine," established in 1823 by the late Mr. Robertson, who many years ago honoured me in that periodical with the title of "Father of this important and wonderfully increasing branch of the Arts."

Although duly appreciating the impermeability of Caoutchouc, it is its elastic property that is so valuable, — a property which no other substance possesses in the same degree or kind, and hitherto

PREFACE.

nothing has been discovered which would even be a substitute for it; and yet, though so unique in its character, it is not unfrequently applied to purposes where neither its impermeability nor elasticity are required. There are exceptions to every rule, and for some purposes rubber vulcanized to the hard horny state may be advantageously employed; but the article should be used in this form only for special purposes. There are other materials of a non-elastic nature in abundance, the comparative cheapness of which is the best safeguard against the perversion of so valuable a substance as rubber from its legitimate use, and which will secure it to the manufacturer at moderate cost for the uses for which, from its peculiar properties, it is so admirably adapted.

In public discussions, the word Caoutchouc has been objected to, as being difficult to pronounce, and for that reason it should be called "*Rubber.*" I have also adopted this name, which is, indeed, its ordinary one.

I have inserted all the authentic statistical Tables I have been able to procure.

I have also subjoined a list of most of the articles we at present commonly manufacture, and a few engravings to illustrate some of them. I do so for the information of such of my readers who might not know the extent to which this manufacture has been carried, and also as a record of what has been done in our time for the amusement of those who are to succeed us.

In this place, as an act of justice to my late partner Mr. Macintosh, I may quote a passage from a memoir (printed for private circulation) by his son and successor in the firm—the late Mr. George Macintosh—as it furnishes a detail of the circumstances which led to his Invention of the Waterproof Double Textures, that made his name celebrated throughout the world : —

" Upon the introduction of coal gas in Britain for the purposes of lighting apartments, and the streets of towns and cities, the manufacturers of the article found that the tar and other liquid products resulting from the process accumulated upon their hands, in the shape of a most disagreeable and inconvenient nuisance. Mr. Macintosh, chiefly with the view to the production of ammonia to be employed in the manufacture of Cudbear, entered, in 1819, into a contract with the proprietors of the Glasgow gas works, to receive for a term of years the tar and ammoniacal water produced at their works. After the separation of the ammonia in the conversion of the tar into pitch, to suit the purposes of consumers, the essential oil termed

naphtha is produced; and the thought occurred to
him of its being possible to render this also useful,
from its powers as a solvent of caoutchouc, or india
rubber. By exposure to the action of the volatile
oil termed naphtha, obtained from the coal tar, he
converted this substance into a waterproof varnish,
the thickness and consistency of which he could
vary according to the quantity of naphtha which he
employed in the process. Mr. Macintosh obtained
a patent for this process in 1823, and established a
manufactory of waterproof articles, which was,
in the first instance, carried on at Glasgow; but
eventually he formed a partnership concern with
Messrs. Birley and other friends in Manchester,
where operations on a very extensive scale were
entered upon, and the business carried on under
the firm of Charles Macintosh and Co."

I have taken considerable pains whilst writing
my narrative to be correct in its facts, and to this
end I have ransacked letters, papers, and books, and
have also had recourse to old samples kept from
time to time as memorials of my progress and
success in the pursuit to which I have devoted
myself. This search has brought to my remembrance much of the minutiæ which occurred on
different occasions, and under the varied circumstances, wherein I have been called on to act. After

the lapse of thirty years, these had in many cases escaped my memory.

Although the task has been somewhat laborious, I do not regret having undertaken it; and if my reader is young, it may serve to stimulate and encourage him to find that, with very slender means and small beginnings, by care and industry, with the blessing of God, he may eventually hope to reap the reward of his exertions. To those who need no such encouragement, my narrative may prove interesting, as divulging the origin and progress of a new and useful manufacture. Such as it is, I present it to the public, hoping that, being the production of one little versed or skilled in an undertaking of this kind, the utmost indulgence will be extended to the author.

I wish to observe, that although there have been many new applications of Rubber and improvements in those which I have not thought it necessary to particularise, yet I believe there is no really new mode of manufacturing the substance itself beside those I have mentioned. If any such exists, it has not come under my notice.

THOMAS HANCOCK.

Stoke Newington,
21st November, 1856.

ORIGIN AND PROGRESS

OF THE

INDIA-RUBBER MANUFACTURE
IN ENGLAND.

DURING the space of thirty-six years in which I
have been engaged in the manufacture of Caoutchouc or India-rubber, I have frequently had inquiries made both as to the motive which prompted
me to commence my first experiments on this substance, and also the manner in which, during my
progress, the valuable properties it possesses were
developed and applied to so great a variety of
purposes. These inquiries have by no means ceased,
and I know of no way in which I can more readily
reply to them than by carrying into execution a
design I have long had in my mind of writing a
simple narrative of things as they have occurred, as
far as the failing memory of seventy years and the
memoranda I may muster will permit me.

I have no very clear recollection when I first
began to notice the peculiar qualities of India-

B

rubber, but well remember that the more I thought about it and tested its properties, the more I became surprised that a substance possessing such peculiar qualities should have remained so long neglected, and that the only use of it should be that of rubbing out pencil marks. I had spent my earlier days in mechanical pursuits, and was well acquainted with the materials generally employed therein, and also with the use of tools, so far as to enable me to make with my own hands almost any kind of machinery required to carry out my views ; but of chemical knowledge I had almost none. I premise this because it will be seen in the course of my narrative, that, although the substance I was contemplating apparently required to be treated chemically, I owe my success principally to the practical knowledge and the degree of skill I had acquired in mechanical manipulation, which proved eventually to be the best preparation I could have had for operating upon rubber ; and it is a singular fact that, although this substance had attracted the notice of chemists from the earliest date of its importation into Europe, and some of the ablest had employed themselves upon it, they failed to discover any means of manufacturing it into solid masses or to facilitate its solution. I was at first imbued with the notion that, to make it useful, I must find a good solvent ; and I think my first experiments were directed to some attempts to dissolve it in oil of turpentine, but I found I could only make very thin solutions, and these dried so badly, or rather not at all, that they were useless. The oil of turpentine then procurable was no doubt of inferior quality; when pure, it dries perfectly. This was about the year 1819. Failing in making useful solutions, I began to think it might be applied as an elastic to various purposes, particularly to articles of wearing apparel. I knew that, although perfectly flexible and extremely elastic when warm, it became rigid when exposed to a low temperature, but still that the warmth of the body was sufficient; consequently, when in use, it retained its elasticity. After various trials, I entertained no doubt that I could adapt it to many uses where elasticity was desirable.

My first patent was dated the 29th April, 1820, " For an improvement in the application of a certain material to various articles of dress and other articles, that the same may be rendered more elastic." The specification of this patent was settled by Mr Bolland, then at the bar, afterwards Baron Bolland, and will be found in the Appendix. I will therefore only mention here some of the purposes enumerated therein: to the wrists of gloves, to waistcoat backs and waist-bands; to pockets, to prevent their being picked; to trouser and gaiter straps, to braces, to stockings and garters, to riding-belts, to stays; to boots, shoes, clogs, and pattens, when the object is to put them on and off without lacing or tying; to the soles of shoes and boots, &c. &c.

When I began to carry my inventions into practice I found some unexpected difficulties. The

India-rubber springs must of course be attached in some way to the article to be elasticated, and women were set to work to sew them in with needle and thread; but after they had been a short time in use I found the holes made by the needles so many tearing-places, and also that if a needle was passed through any part of the rubber it endangered a fracture. To remedy this in some measure, I made the ends of the springs much thicker and wider than the central parts, so that the weaker part yielded its elasticity sufficiently without bringing much strain upon the thick ends, where the punctures of the needle were made. But the necessity for this form of spring brought other difficulties: instead of cutting in straight lines, each spring had to be hollowed out, tapering on both sides and both edges, and contrivances had to be adopted for this purpose. The needle-holes did not then tear out. I soon found that the knives and cutting tools required to be kept wet with water. These springs, however, had not long been in use before they were returned in numbers broken. By pulling out new springs smartly and allowing them to return quickly, I observed, after a time, that the angles became finely serrated like a saw, and each nick increased in size as I followed up the operation, until the spring snapped in two. This appeared to be a very formidable obstacle to success as regarded springs, but was soon overcome, for I observed that some of the new springs were, and some were not, affected in this way; and on tracing back the steps that had been taken with the two kinds, I found that those springs on which I had used boiling water, after they were cut, did not crack on the edges ; and I had no farther trouble on this score, always taking care in all future cases, when edges were freshly cut, to give them a hot bath.

This discovery was of great value to me, as furnishing the fact, at this early period, of the great importance of employing heat in treating this substance, as will be seen throughout the progress of its future manufacture. At the period of which I am speaking, there was brought into this country a certain proportion of small thin bottles of rubber ; these I selected, and cut them into rings. The rings, requiring no sewing, were used chiefly for the wrists of gloves ; they had the hot bath, and were then passed on to the glove, and a strip of thin leather sewed over them, so that they were in a kind of pipe. Springs were put into stockings in the same way ; both the stocking or glove and spring being kept to their utmost tension, whilst the leather slip, or tape, or ribbon, was stitched on, and the rubber, when set at liberty and warmed, gathered up the whole by its resilient action. It will be seen by referring to the specification the extent to which this patent has been applied; and although there have been great improvements made since in these applications (as will be seen hereafter), yet the principle is the same — that is, the acquiring of any degree of elasticity by means of a soft flexible substance in articles of dress, which before was only to be acquired by means of hard springs of steel or wire.

I also applied rubber to the soles of shoes and boots both externally and internally, which rendered them both elastic and waterproof; but this was done at first with only such rubber as I found thin enough amongst the raw material, and this I flattened by heat and pressure.

It was not long before the waste cuttings of the rubber began to accumulate fast; and with the then scanty supply in the market, particularly of the kind fit for my use, I foresaw that I should soon be at a loss for suitable material, unless I could find some means of working up the waste, and I at length resolved upon attempting it. I remember that, in my experiments about this time, I employed a Papin's digester, such as is used for culinary purposes, but the only result that I recollect was the production of a thick fluid of the appearance of treacle: I could, however, make little or no use of it.

I had observed that when pieces with fresh-cut edges had been long in the hot bath, and then dried, they would perfectly unite; but the outer surfaces, which had been exposed, would not unite, however clean, or however heated or pressed. This was the more perplexing, as the rubber came to this country in irregular shapes and forms, rendering almost impossible, at any reasonable cost, the paring of the surfaces. However, I resolved to make a beginning (for the want of which beginning we often fail of things within our reach). My first step was to procure a hollow punch one inch square ; with this I punched out squares of rubber. I chose this small size that I might waste the less in facing them. In the meantime I got an iron mould made of the size of the squares : into this mould, which was perfectly true inside, I had a plunger fitted at each end, and, putting the surfaced squares into the mould and then the plungers, I submitted them to severe pressure in boiling water. On withdrawing my charge, I discovered that air had got between some of the surfaces and prevented a perfect union, but others were quite perfect. I now found that I could thus obtain solid blocks, four or five inches long, and one inch square, and true in form. This was a grand move, as it enabled me to make another step in advance, namely, to cut from the surface of the end of the blocks sheets of any thickness. This I did by means of a circular knife in a lathe, the block being confined in a trough, having a screw at the end which would by a turn bring up the block to the knife in succession as the sheets were cut off, the trough and block being carried past the knife by a slide motion : the knife working in water, and having a keen edge, I cut off sheets exceedingly thin and smooth, well suited to some of my purposes. I have a good part of one of these blocks now before me.

Although I had thus advanced one step in forming solid blocks, still this could only be done with the thickest and best of the bottle-rubber ; and did

not aid me at all in using up my waste cuttings.
Finding that fresh-cut surfaces united so perfectly,
I began to consider how I could improve upon
this knowledge in uniting smaller pieces; and it
occurred to me that, if minced up very small, the
amount of fresh-cut surface would be greatly increased, and by heat and pressure might possibly
unite sufficiently for some purposes. I accordingly put this plan into operation; but although
I spent much time and took great pains to bring
about a good result, I could not succeed to any
useful extent. The uncut surfaces, however small,
would not unite either to each other, or to a cut
surface, so that the mass easily separated, and I
was obliged to abandon this mode.

These discouragements were for a time very
vexatious, as my means were but slender. Although I was making way, I could perceive that
unless some mode could be found to unite not only
the waste cuttings, but also a large proportion of
the material as imported (which was so uncouth
in form, and irregular in surface and size, that it
could at present be turned to no useful account),
my object would not be attained. My mind being
solely directed to this subject, I saw the prospect
of new applications to an enormous extent of a
substance with the properties of which I was daily
becoming more and more conversant. I did not
give up the pursuit: the object I had in view
seemed within my reach by what I had already
done, but the object itself I could not yet grasp.
Revolving in my mind the readiness with which

newly cut surfaces would unite, I thought that a tearing action might do better than simply cutting. This could only be done by a machine, and I accordingly constructed a small experimental one, such as I thought most likely to effect the tearing of the rubber into small shreds. I can best describe this by a sectional sketch. (It is given in perspective, with the large masticator.)

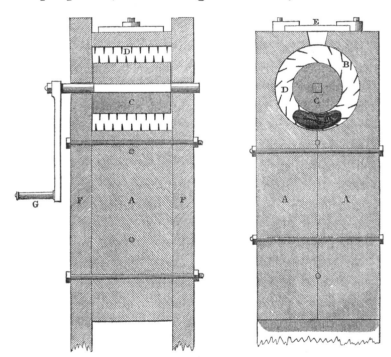

A A, two pieces of wood bolted together.
B, a hollow cylinder cut out of A A, and studded with teeth.
C, a cylinder of wood studded with teeth, and having a spindle passed
    through it.
D, space between the two cylinders B and C.
E, an opening with a cover.
F F, two pieces of wood bolted on both sides of A A, and enclosing the
    space D, and cylinder C.
G, a winch.
The darkened spot in space D represents the charge of rubber.

At the top of the hollow cylinder is an opening, into which was put some hot rubber : when closed, the cylinder C was put in motion by the winch G. The rubber being now dragged in, and the motion continued, the teeth began to operate, and it soon became evident that some action was going on inside that I had not reckoned upon, as much greater power became necessary to turn the winch. After turning some time, the hole at the top of the hollow cylinder was opened, and presently, to my great surprise, came out a round solid ball. This ball, when cut open, presented a marbled or grained appearance ; the union of the pieces was complete ; the graining exhibited the pieces curiously joined together, the exterior surface of them having been acted upon so as apparently to alter their condition, whilst the interior portion of the pieces seemed to be in the same condition as when put in. The ball was replaced and the action was continued for a long time, and when taken out again it had become very hot ; and on cutting it open all the graining had disappeared ; the whole had become a solid homogeneous mass. This operation was repeated until my experimental machine, constructed chiefly of wood, would no longer hold together, and I lost no time in applying to Messrs. Hague and Topham, the engineers, to make a proper working apparatus upon the same principle. With my experimental machine I could not operate on more than about two ounces of rubber for a charge, and I found this quantity required nearly

the power of one man to work it. I therefore calculated the capacity of my new machine for one
pound, and, in order to enable one man still to act
upon it, I had the speed reduced by one-half by
spur-wheel gearing. I had it made very strong, as
I found the charge might be increased to any
amount that the space between the cylinders would
admit, provided sufficient power was applied to
give motion to the cylinder. Experience taught
me afterwards that, with a smaller charge and increased speed, I could produce the same effect; in
other words the result depended entirely upon the
amount of motive power employed, and whether
fast or slow.

Whilst this machine was being made, I reflected
on the effect produced on the rubber by this
singular operation. I observed that if the rubber
was put into the machine hot and dry, the effect
was only thereby hastened a little; for, if put in
cold, it soon became heated: nor was it of much
importance whether the pieces of rubber were
larger or smaller, so that they were dry. The pieces
soon began to unite, and presently all were worked
up into a rough uncouth shape; but by continuing
the action, the roughness and deformity gradually
disappeared, until at length the mass assumed
a regular spherical form, the exterior merely
showing the indentations made by the teeth.

I deduced from these facts that the union and
consolidation of scraps, cuttings, bottles, shoes, or
lumps of rubber, promiscuously thrown into this

machine, was due to the combined action of heat and motion under severe pressure: when a heavy charge is operated on, the heat it acquires is very surprising. I have since found, on cutting a heavy charge open and closing it upon the bulb of a thermometer, that the temperature reached 280° Fahr., and this heat could only be due to the motion of the machine and the action of the rubber upon itself during the transition state, as the same resulting temperature was attained when the rubber was put in cold and the machine also cold.

My new machine was at length delivered, and I found my calculation correct; a man could just manage to keep up the action of the cylinder with a pound charge of rubber in it. The machine wrought the charge into a cylindrical form, which it assumed in a very short time, and then evidently revolved upon its own axis around the solid cylinder: the charge came out, I think, about seven inches long, and one inch and a quarter in diameter. I had now at command the means of reducing all kinds of rubber, whatever size or form the original pieces might be, to a solid mass, without any foreign admixture, or the use of any solvent, or having recourse to any chemical process, the effect being produced solely by a mechanical action on the rubber itself disturbing the original structure of the substance and recomposing it, without materially altering its peculiar qualities, or unfitting it for any of the purposes to which it could be applied in its naturally constituted state.

I wish here to remark that the discovery of this process was unquestionably the origin and commencement of the india-rubber manufacture, properly so called : nothing that had been done before had amounted to a manufacture of this substance, but consisted merely in experimental attempts to dissolve it; and even this had never yet been effected for any useful purposes. These experiments and the results produced occurred during the summer of 1820.

The manufacturing of promiscuous forms of the raw material into solid masses and combinations without the use of solvents, now called masticating, has been from time to time adopted by others, and even introduced into specifications without the least acknowledgment, and not unfrequently quoted from these into scientific works, as if originated by the parties who had taken this liberty and adopted it as if it were their own.

I have mentioned that the new machine produced the rubber in a cylindrical form ; but this form did not suit me, and I then had recourse to an iron mould, which would exactly contain one charge of the machine ; and as the charge immediately on removal was in a semi-plastic state, pressure quickly applied caused it to conform to the shape and size of the mould. I have preserved one of these first small blocks as a curiosity ; it is now lying before me ; it was the produce of a very small charge, and measures six inches long, one and a half inch wide, and three-quarters of an inch thick. Although I could cut some of my elastics and other things from these small blocks, yet I soon found it necessary to increase their size; and whilst a larger machine was being constructed I had moulds made of sufficient capacity to contain four of the largest charges the present machine would produce, and, putting them together into the mould whilst hot, they united perfectly, and gave me a block four times larger.

I remember at that time, when exhibiting a piece of my solid rubber to an old gentleman, he examined it, and on returning it made this remark (which bids fair to be realised): "*The child is yet unborn who will see the end of that.*"

In the place I then occupied I could only employ manual labour; my second new machine was calculated for the power of two men, and by uniting four charges of this machine I obtained blocks of considerable size; but the demand increasing, I found it necessary in the following year, 1821, to move into larger premises in Goswell Mews, Goswell Road, London. Here I had a horse-mill put up, and connected the power not only to larger machines, but also to iron rollers, which I now found very useful; as, by passing the raw rubber through them several times when hot, it formed itself into a kind of rough corrugated sheet, which not only brought it into a good state of preparation for the machine, but greatly facilitated the drying,

—a process which became more and more necessary, as I made purchases of newly imported rubber, and therefore frequently in a moist state.

I think my blocks now amounted to fourteen or fifteen pounds each, and all the operations could be carried through with certainty and despatch. I must not omit to state here that, all through these operations, heat was indispensable; and, when it became necessary (as it soon did) to hasten the production, it was found of great advantage to expose the rubber before it entered the machine to as high a temperature as could be safely adopted. To carry this work out, I had a brick oven built, and employed a regular baker to attend it. The rubber was placed in earthern pans, and he was directed to keep his oven at such a heat as would make the rubber as hot as possible without melting it : and this he did with tolerable accuracy, judging of the temperature by modes at that time used by bakers, to which he adhered; but as he did not use a thermometer, I cannot now give the degree of heat. I have no doubt it was sometimes as high as 300°. This mode of heating was followed until, in 1822, I began to heat the rubber in metal vessels surrounded by high-pressure steam. During the process of mastication I sometimes introduced colouring matters : they combined perfectly with the rubber : the colours were not very good, as the dark colour of the rubber injured them.

As I took no patent for my process, it was of course an object with me to keep it secret. I

pledged my men to this, and treated them well;
and they in return kept faith with me, and, in order
to disguise the matter as much as possible, the
machine was called " a pickle," and retained that
name long after the secret became public: it
has since been called a masticator or masticating
machine. This name having now become common,
I shall hereafter use that designation. Whilst on
this subject, I may just mention that I kept this
process perfectly secret for twelve or thirteen
years, that is till about the year 1832.

I believe I may now dismiss the description of
the masticator with this remark, that no alteration
has been made in the principle of its construction
or use, except the omission of teeth in the hollow
cylinder, as it was soon found that the mass would
revolve without them. Of course the dimensions
have, step by step, undergone great changes. The
first charge ever produced did not exceed two
ounces; and the masticators now in use at our
works in Manchester are charged with from 180 to
200 pounds each; and the blocks resulting, without
joining, are six feet long, twelve or thirteen inches
wide, and about seven inches thick.

At first, in 1820, some of the blocks were cut into
springs, and used for other purposes of my patent,
but very soon after into square pieces, and sold by
stationers for rubbing out pencil marks, much of
the present form and size. In a short time it
became obvious that the blocks might easily be cut
into sheets and used for a variety of purposes ; and

to effect this object, a simple machine was constructed. It consisted of a wooden box of the size of the largest blocks I then could make, probably about eight inches long, four wide, and three thick. This box had a movable bottom, which was raised or depressed by four long screws, acting in the fixed bottom of the box; the upper ends of the screws, having a shoulder, were made to act in metal holes, on the lower side of the movable bottom; on the upper edge of the sides of the box, smooth steel plates were attached. When the machine was to be brought into operation, the screws were drawn quite down, followed by the movable bottom, and the block of rubber with its sides well soaped, was put into the box and pressed down upon it; the screws, having cross handles at the lower end, were then equally turned until the block of rubber rose a little above the steel plates on the upper edges of the box; a strong, straight knife with a keen edge, kept wet by water dripping upon it, was then inserted at the right-hand end, and by a steady cutting and thrusting motion passed through the block to the other end. This first cut took off the rough surface, and this was continued until a smooth, solid surface was obtained; then followed the cutting of sheets by giving each of the screws as many turns as were necessary to raise the block above the surface of the steel plates to the thickness of the sheet required; and then repeating the operation with the knife, a beautiful clear sheet of rubber was produced, and could by this machine be cut so thin as to be semi-transparent. The sheets when warm could be joined edge to edge with great facility; and large sheets were made in that way.

During the early part of 1822, a person who had a patent for uniting two pieces of young cork together, with the intention of producing an article more free from the perforations in ordinary cork, applied to me to cover with rubber so much of the cork as projected above the neck of the bottle, to preserve the cement with which they were united from the effects of mildew. I mention this as one of the first applications of the cut sheet rubber, and as having been the cause of turning my attention again to discover some practical mode of producing a useful solution of rubber to effect the capping of these corks; and I was agreeably surprised to find that my manufactured sheet rubber yielded to good oil of turpentine with the greatest facility. I found, however, on applying it to the corks, that, although the solution dried perfectly, the corks when covered with it, and afterwards brought in contact, would unite together: this adhesive character of the solution when dry afterwards became of great value, not only for these corks, but in a great variety of ways, and continues so to this day. I ultimately applied the cut sheet rubber to the capping of the corks; and as this was done in a novel manner, I wish to show how gradually the qualities and uses of rubber became developed, and the step-by-step progress that was first made in adapting it to practical purposes. A great number of boys were employed about this work; but I will proceed to describe it in the person of one. He had a small kind of lathe, with multiplying gear, turned by his left hand; there was a taper hollow chuck to the lathe, of the size to receive corks of the varying diameters; he had also a small rest for his hand, and a small pot of rubber solution (now called varnish); he put the cork into the chuck, and slightly adjusted it; and then, dipping his brush in the rubber-varnish, he laid it on the cork at the proper distance onwards from the end, and turning the lathe, the intended portion of the side of the cork, and its end was covered. When dry, it was ready for the sheet capping, which was effected in the following manner : — The pieces of sheet rubber employed were about an inch square, and something less than one sixteenth of an inch thick; these were laid on a flat tin vessel heated by steam; the boy had a contrivance, fixed to his bench, which held the cork with the varnished end upwards; the whole turning on a pivot, so that the cork could be turned round. The cork being so placed, the boy took the square of rubber hot from the tin vessel, and dexterously and quickly extending it in all directions with his fingers, until he more than doubled its dimensions, clapped it in an instant on the top of the cork, and pulling it down over the sides of the cork, at the same

moment twirling it round between his hands, the
cap was secured to the cork as far as he had before
laid on the varnish ; then passing a knife sharply
round, the superfluous sheet flew off, leaving the
cork beautifully covered with a neat rubber cap.

I have mentioned that I had found solutions of
rubber of considerable consistence could be made
with the manufactured rubber with great facility ;
this I attribute to its having undergone a previous
disintegration in the masticator, whereby it is
also somewhat softened, admitting a more ready
penetration (if I may use that expression) by the
solvent ; at all events, it is a fact that masticated
rubber dissolves freely.

Finding that pitch and tar could be readily
combined with rubber, in the course of 1823 I
obtained a patent for these compounds. I at
first mixed the tar with stiff solutions of rubber,
and the pitch likewise, by melting the latter,
and treating the compound with heat, both during
the mixing and when using it ; and these are
the modes stated in my specification of that
patent inserted in the Appendix. I afterwards
found in practice that pitch would combine
readily with rubber in the masticator, and thus
save the cost of solvent. This compound was
made of different proportions of the materials ; the
two most useful being equal weights of each, or one
of rubber to two of pitch. It was necessary to use
water in the masticator to prevent adhesion to the
cylinder. The facility with which these substances

combine, and the prodigious increase in bulk whilst hot, were surprising.

These compounds were spread by means of hot iron rollers, and a wet cloth passed with the compound rubber between the rollers. I thus obtained a smooth sheet of equal substance throughout, and of any required thickness. Mr. Cassell of Poplar, who supplied me with coal naphtha, was at this time applying coal tar asphalte of different degrees of hardness to road-making and other purposes; and I obtained from him samples of this material, which I combined with the rubber in the masticator, as I had done the vegetable Stockholm pitch, but I found the latter to combine more readily, and generally preferred it. Claims have recently been made to these compounds of rubber, pitch, and asphalte, but they are not only included in this patent, but also in my patent of 1843.

Sheets of this compound were applied extensively to the sheathing of ships bottoms, under the copper, as a protection against the destructive ravages of worms, which do immense damage to the timbers in some waters; and for which sheets of tarred felt had hitherto been used. I remember that in some instances sulphur, cow's-hair, and other things, were mixed with the compounds of rubber and pitch used for this purpose, as substances likely to repel the worms; these compound sheets were submitted to pressure between hot plates. The first vessel sheathed with the sheets was the yacht of the first Sir W. Curtis, and the second was the Kinnersley

Castle. The specification of this patent was settled in consultation by Sir John Copley, the present Lord Lyndhurst, and duly enrolled. I subsequently parted with this patent to my late brother Walter Hancock, and others, who, after making great improvements in the mode of expediting the production, and sheathing a great number of ships, got into litigation amongst themselves, and the business ceased.

In the year 1823, one of my late partners, Mr. Charles Macintosh of Glasgow, obtained a patent for rendering two fabrics waterproof, uniting them with a solution of rubber ; hence they were called " waterproof double textures," and afterwards came to be universally known by the name of " Macintoshes."

Early in the year 1825, I obtained a license from Mr. Macintosh for the use of his patent ; he had also entered into engagements with parties in Manchester to carry out his plans, and a large building, with machinery, was erected there for the purpose. Mr. Macintosh manufactured the varnish at Glasgow, where he also manufactured the coal naphtha which he used as the solvent of the rubber. I had from the commencement some advantage over the firm at Manchester, inasmuch as my concern was not only already in operation, but was also on a more limited scale, and therefore could be pushed on with greater despatch. At length, however, the two concerns were moving, each in its sphere. Mr. Macintosh's solutions were very thin, and therefore penetrated more into the textures; and consequently the odour of the naphtha was not only very prevalent when they left the works, but they also retained this strong odour for a very long time. In this respect I had a decided advantage. I employed for my solvent equal parts of naphtha, and very pure oil of turpentine, which greatly mitigated the smell of the goods. But I had another and still greater advantage, inasmuch as my solutions were made with masticated rubber, and consequently with less than one half the proportion of solvent which Mr. Macintosh found it necessary to employ in the process by which his solutions were made. Knowing this to be the case, I wrote to Mr. Macintosh, and offered to supply Messrs. Macintosh and Co. with my solution; but this offer was at that time declined. In the meantime these fabrics were quietly becoming known to the public, and the goods were taken up nearly as fast as we could respectively produce them.

The late lamented Captain John Franklin (afterwards Sir John Franklin), in a letter to Mr. Macintosh dated 30th April, 1824, after acknowledging the receipt of a large quantity of waterproof canvas for covering boats, &c., says, "Will you also make up four life preservers of a size for stout men, and eighteen bags about six feet long, and three broad, fitted with corks for filling with air for the party to sleep on, and four for pillows of the size of the one you gave me." I insert this extract to show how early these waterproof double textures were appreciated, and the application of the material to air-beds, pillows, and life-preservers.

With the view of keeping up something like a chronological order in my narrative, I will for the present leave the double textures, and resume the subject of the progress made with my first patent of 1820.   It will be remembered that, in applying the rubber springs, considerable difficulty was experienced in attaching them securely to the articles to which they were otherwise so well adapted. This difficulty was now entirely removed.   The facility with which solutions could be made of a consistence suitable to the purposes of a cement suggested a new mode of making elastics, which has continued in use to the present time.   In order to illustrate this new mode, I will take the instance of a glove.   The glove was turned inside out, and a plate of tin thrust into the top of it, of such a size as to stretch the glove to its greatest width; a coat of solution was then laid either on one side of the glove or all round it, but generally on one side only, about half an inch wide, and then laid aside to dry.   As soon as a sufficient number were done to admit of the first glove becoming dry, it was then ready for the next operation.   A small light wooden frame was prepared about a foot long, having a piece of tin attached to each end; these pieces had very narrow slits cut in them, about an eighth of an inch apart; the rubber spring, now called rubber thread, cut to a proper size and length, and taken from a hot plate, was stretched the whole length of the frame, and the ends entered into the slits; for gloves there were generally three threads used. Three or four gloves in the position before described were laid side by side, and the extended threads in the frame placed on the prepared part of the glove, and pressed down upon them: a slip of kid leather or silk cut diagonally (to admit of its more ready contraction) coated with solution was then placed over the threads, and rubbed down.   The threads were then cut off at each end, when the tins being taken out of the gloves, the resilient action of the rubber immediately contracted the glove to nearly its original size, neatly corrugating the leather. This was a great improvement, as no sewing whatever was necessary, and the top of the glove rendered perfectly elastic, so as to admit the hand with ease, and then contract to the size of the wrist.   An immense quantity of gloves were thus elasticated until another improvement was introduced some years after, of which more anon.

By this mode a great number of articles were rendered elastic: such as garters, braces, trouser-straps, shoe gussets, waist-bands, knee-caps, bands, and for a variety of surgical and other purposes. This was the origin of what has since been called " shurred " or corrugated goods; some of the manu-facturers following precisely this mode, others have introduced machinery, but all their goods are made on the same principle.

The application of the thick solution enabled me

to render the leather used for the soles of shoes and boots perfectly waterproof. This was done by coating the leather first with solution, and applying when warm a thin sheet of the rubber, cut from the masticated blocks. Upper leathers of shoes and boots were also rendered waterproof by a novel mode of operation. I have before mentioned that these solutions continued adhesive after they were perfectly dry. Adhesiveness of the surface was in this case objectionable, and it was remedied thus. The leather was coated with solution, and laid aside to dry; a piece of the sheet rubber cut from blocks, and about four or five inches square, was laid on a hot plate; a piece of tin of the required size, say nine or ten inches square, with the edges turned down all round and serrated, was fastened to a block of wood about one inch smaller all round than the tin; the sheet rubber was then taken from the hot plate and stretched out skilfully with the hand, and brought over the serrated edge of one side of the tin, and then carefully stretched over the other three sides, the serrated part holding the sheet in its extended state, whilst the leather, coated with solution, was laid on and well rubbed down; the edges of the rubber were then cut round, liberating the leather, covered with a beautiful, thin, smooth, unadhesive film of undissolved rubber. Both upper leathers, quarters, and soles were thus done in considerable quantities.

Drawing masters and others using black-lead pencils approved highly of the neat square blocks cut from the masticated rubber, and the quantities in demand for this purpose constantly increased ; they are still supplied to a large extent. Pieces, blocks, and forms of much larger dimensions, began to be inquired for, and made for purposes with which I was not made acquainted. The edges of wheels and the surfaces of rollers and cylinders were also covered with rubber of various thicknesses for machinists. The billiard-table makers also applied for long, evenly cut pieces to form the cushions of their tables, which were successfully applied, and have continued to be used for that purpose (with modifications) ever since, to the exclusion of all others. In the early part of 1822, I began to make tubing of the cut sheet rubber, and afterwards with alternate plies of cloth coated with solution ; some also were covered with leather, velvet, &c. I had forgotten many of these things till I had recourse to my books and memo-randa; many of which, on making search, I have unexpectedly found, with specimens made at the time, also preserved.

In the course of my early progress, I found that some of the rubber I employed was very quickly decomposed when exposed to the sun. As the heat was never more than about 90°, and rubber when exposed to artificial heat of a much higher tem-perature was not injured by it, I suspected that light had some effect in producing this mischief. To ascertain this I cut two squares from a piece of white rubber; one of these I coloured black, and

exposed the two to the sun's rays ; in a short time
the piece that had been left white wasted away, and
the sharp angles disappeared, and it assumed the
shape of a piece of soap that has been some time in
use ; the blackened piece was not at all altered or
affected. The lesson taught me by this experiment
was of great value ever after.

The following extract will show how early the
masticated cut sheet rubber came under the notice
of scientific persons. The sheets were supplied to
Professor Faraday.

THE QUARTERLY JOURNAL OF SCIENCE, LITERATURE,
    AND THE FINE ARTS.   Edited at the Royal Insti-
    tution of Great Britain, vol. xvii.   Lond. 1824 :
    p. 364. No. XXXIV. : —
" Mr. T. Hancock has succeeded, by some process the results of long investigation, but which
he has not published, in working caoutchouc with
great facility and readiness. It is cast, as we
understand, into large ingots or cakes, and being
cut with a wet knife into leaves or sheets about
one-eighth or one-tenth of an inch in thickness, can
be applied to almost any purpose for which the
properties of the material render it fit. The caoutchouc thus prepared is more flexible and adhesive
than that which is generally found in the shops,
and is worked with singular facility. Recent
sections, made with a sharp knife or scissors, when
brought together and prepared, adhere so firmly as
to resist rupture as strongly as any other part, so
that if two sheets be laid together and cut round,

the mere act of cutting joins the edges, and a little pressure on them makes a perfect bag of one piece of substance. The adhesion of the substance in those parts where it is not required is entirely prevented by rubbing them with a little flour or other substance in fine powder. In this way flexible tubes, catheters, &c. are prepared; the tubes, being intended for experiments on gases, and where occasion might require they should sustain considerable internal pressure, are made double, and have a piece of twine twisted spirally round between the two. This, therefore, is imbedded in the caoutchouc, and at the same time that it allows of any extension in the length of the tube, prevents its expanding laterally.

" The caoutchouc is, in this state, exceedingly elastic. Bags made of it, as before described, have been expanded by having air forced into them, until the caoutchouc was quite transparent, and when expanded by hydrogen they were so light as to form balloons with considerable ascending power, but the hydrogen gradually escaped, perhaps through the pores of this thin film of caoutchouc. On expanding the bags in this way, the junctions yielded like the other parts, and ultimately almost disappeared.

" When cut thin or when extended, this substance forms excellent washers or collars for stop-cocks; very little pressure being sufficient to render them perfectly tight. Leather has also been coated on one surface with the caoutchouc, and without being at all adhesive, or having any particular odour, is perfectly watertight.

" Before caoutchouc was thus worked it was often observed how many uses it might in such a case be applied to; now that it is so worked, it is surprising how few the cases are in which persons are induced to use it. Even for bougies and catheters it does not come into use, although one would suppose that the material was eminently fitted for the construction of those instruments."

I must now enter upon quite a new era in the manufacture of rubber. In the month of May, 1824, I received from Central America (I believe Guatamala) a considerable quantity of this substance in the pure liquid state as drawn from the trees; it came in vessels formed of two or three joints of the larger kind of cane, four or five inches in diameter, with a small hole at the end securely stopped. This liquid was of the consistence of thick cream, and of exactly the same colour, so that in appearance it might have been taken to be cream. I found that when the moisture was evaporated from it by exposure to the atmosphere, the residue was pure rubber of the finest quality, but it had lost more than one half its weight when in the liquid state. I made many experiments with it, and found little difficulty in applying it to various useful purposes, and obtained a patent for one of these objects, which was sealed on the 29th Nov., 1824. I was assisted in writing this specification by the late Mr. Bryan Donkin, C.E., and it was afterwards settled in consultation by Mr. Frederick Pollock, the present Lord Chief Baron, and duly enrolled, and will be found in the Appendix.

The principal feature in this patent is the manufacture of a kind of artificial leather, which was produced by saturating felt, carded cotton wool, and hair, and in combining other fibrous substances, such as hemp and flax, with the liquid rubber, and when dry submitting the whole to pressure: by these means a very strong, tough, and useful material could be made, very much of the appearance of real leather, and of various degrees of quality, some suitable for rough purposes, such as soles of shoes, hose-pipes, straps, harness, &c., and others very thin, soft, and flexible, and of every variety of colour, all of them having this advantage over real leather, in being more or less waterproof according to the quantity of liquid rubber employed in more closely or loosely uniting the fibres. The surface when exposed to the atmosphere lost its adhesiveness; and as colouring matters could be freely mixed with the liquid, the external surface of the articles could be treated with such coloured liquid so as to cover the fibres, and leave a smooth face of any required colour. The liquid when dry became of a dark colour, but when necessary the colouring matter could be removed by repeated washings in clear water: the colouring matter subsided with the water, when the clear liquid floating on the top could be drawn off: this washing could be carried on until the resulting clear liquid rubber, when dried, was colourless, and rendered fit to receive any delicate tint.

Some authors have stated that the natives, when making articles from this liquid, suspend them over the smoke of some particular vegetable for the purpose of drying them, and that this smoking produces the dark colour. From such experience as I have had, I am inclined to doubt whether the colour is so acquired. I found that the liquid when used in its unwashed and primitive state always became gradually of a dark colour whilst drying, and that by exposure to the sun and atmosphere the surface acquired that beautiful soft feel peculiar to a well-finished native article. I have always considered the dark colouring matter it contains to be a kind of natural protection against the effects of light, as I found when deprived of this colouring matter by washing that it speedily became decomposed by the ordinary light of my room. When the first coating has become dry I have found that a succeeding coat would not very readily spread itself equally on the previous surface (as if it were greasy), and I am inclined to think that the smoking alluded to is employed to dry the liquid, and partly to promote a more ready and equal flowing and adhesion of the succeeding coat.

In the following year, 1825, I obtained a patent

for other modes of manufacturing artificial leather, in which I employed a solution of rubber instead of the liquid of the former patent, for combining fibrous substances in their manufactured or un-manufactured state, such as wool, silk, cotton, hemp, flax, or hair, carded, hackled, or felted. A fleece of carded cotton was brought from the carding engine upon a piece of cotton-cloth pre-viously coated with rubber solution before it was quite dry, and then another cloth similarly coated was laid upon the fleece of cotton; the whole was then submitted to pressure so as to force the carded cotton through the solution, which united the fibres into a sheet. The cloth, if previously sized, could be stripped off, and the compound passed between heated rollers, producing a thin paper-like material. Any number could be united into one sheet, of any thickness or length. This prin-ciple of uniting fibrous substances by means of a solution of rubber has served as the groundwork of several patents wherein the invention is claimed as new; but they contain only modifications and mechanical contrivances; the principle is the same. I also united as many plies of cotton, linen, or woollen cloth together as would make up the thickness I required. I also introduced into the solution, for some of the most common purposes, black resin, size, glue, ochre, powdered pumice, whiting, &c. That made of a number of plies of cloth was used during this year, 1825, for the backs of cards in carding machines, instead of

D

leather, and has been found to answer the purpose so satisfactorily that it is now very extensively employed in preference to leather : the advantage being that, as great lengths are required, this material can be made of any length without joining, whereas leather can only be had of the length of the skin or hide. Uniformity of thickness is also essential, and leather requires skilful manipulation and great labour to accomplish this : nor is uniformity of texture and elasticity of less importance. All these conditions are met in this artificial leather. This article is also used extensively as printer's blankets in calico-printing. I also made in this manner very strong straps for driving machinery, and furnished one of the first to the late Sir Isambert Brunel, for his engine, when sinking the shaft for the Thames Tunnel. Such straps have continued ever since, and are still, much used. It was also during this year that I began to make deckle straps for paper-machines, suggested by Mr. Bryan Donkin, C. E. ; and these also continue to be employed for that purpose.

In this same year, 1825, I took out another patent for employing the liquid rubber in the manufacture of ropes and cordage, and other similar articles, with the view of rendering them waterproof. I proposed to use the liquid rubber in the same manner as is commonly practised with tar.

I insert another extract from the Journal of Science, for the purpose of giving to my readers the analysis of pure caoutchouc.

JOURNAL OF SCIENCE, Vol. XXI. pp. 130, 131.

Proceedings of the Royal Institution, London,
February 3rd, 1826.

" The members held their first weekly meeting
at half-past eight o'clock. In the lecture-room
were exhibited a great variety of specimens of
caoutchouc or elastic gum in all its states, from
the uncoagulated crude sap of the tree to that
of perfect purity and aggregation, and also as
united to various fabrics, producing a variety of
strong, flexible, and perfectly water-tight materials,
some being of extreme delicacy, and others of
great thickness and strength. These were furnished for the occasion by Mr. Thomas Hancock,
who has had peculiar opportunities of manipu -
lating with this substance, and possesses the knowledge of a process by which it can be rendered
fluid, and yet retain the power of hardening and
assuming its elastic state again. Mr. Faraday ex-
plained the nature of caoutchouc, and gave the
results of an analysis of the unchanged sap. The
various specimens of cotton, silk, linen, felt, woollen,
&c., which were upon the table, had been rendered
water-tight by the intervention of a layer of caoutchouc between two layers of the fabric,—as for
instance, cotton or silk,—and the adhesion was so
perfect that the substance seemed but as one web.
The perfect retention of water by these substances
was shown by a calico bag, into which a quart of
water had been introduced, and the opening closed

up: not a drop or particle of moisture could be perceived on the exterior, though the bag was much handled and pressed.

"When several folds of calico, linen, or canvas, were cemented together by this substance, a material was produced answering many of the purposes of leather, and surpassing it in value in numerous applications. Its use in the construction of connecting bands for machinery and card-fillets for carding engines has been tried and approved of. In consequence of the manner in which the caoutchouc is applied, no limit occurs as to the form, or size, or delicacy, or strength of the water-vessels or things which may be made : it is equally applicable to the cloak and the caravan cover, to the most ornamented flower-vase, and the strongest water-bucket.

ON PURE CAOUTCHOUC.

QUARTERLY JOURNAL OF SCIENCE AND THE ARTS. ROYAL INSTITUTION OF GREAT BRITAIN. Vol. XXI. No. XLI. London: J. Murray, 1826.

" On Pure Caoutchouc and the Substances by which it is accompanied in the state of Sap, or Juice. By M. FARADAY, F.R.S., Corresp. Memb. Inst. France.

[Communicated by the Author.]

" I have had an opportunity latterly, through the kindness of Mr. Thomas Hancock, of examining the chemical properties of caoutchouc in its pure form, as well as of ascertaining the nature and proportions of the other substances with which it is mixed, when it exudes as sap or juice from the tree. At present much importance attaches to this substance, in consequence of its many peculiar and excellent qualities, and its increasing application to useful purposes. I have thought, therefore, that a correct account of its chemical nature would possess some interest.

" The extensive uses, both domestic and scientific, to which Mr. Hancock has applied common caoutchouc, in consequence of his peculiar mode of liquefying it, are well known. Hence he was fully alive to the importance of its applications when in its original state of division. When he gave me the substance, he communicated many of his observations upon it, which, with others of my own, form the present paper.

" The fluid, I understood, had been obtained from the southern part of Mexico, and was very nearly in the state in which it came from the tree; it had been altered simply by the formation of a slight film of solid caoutchouc on the surface of the cork which closed the bottle. The caoutchouc thus removed was not a five-hundredth part of the whole. The fluid was a pale yellow, thick, creamy-looking substance, of uniform consistency. It had a disagreeable acescent odour, something resembling that of putrescent milk; its specific gravity was 1011·74. When exposed to the air in thin films it soon dried, losing weight, and leaving caoutchouc of the usual appearance and colour, and very tough and elastic : 202·4 grains of the liquid dried in a Wedgewood basin 100° Fahr. became in a few days 94·4 grains, and the solid piece formed being then removed from the capsule, and exposed on all sides to the air until quite dry, became 91 grains ; hence 100 parts of sap left nearly 45 of solid matter.

" Heat caused immediate coagulation of the sap, the caoutchouc separating in the solid form, and leaving an aqueous solution of the other substances existing with it in its first state.

" Alcohol poured into the sap in a sufficient quantity, caused a coagulum and a precipitate, both of which were caoutchouc of considerable purity. The alcohol retained in solution the extraneous matters, which, possessing peculiar properties, will be hereafter described.

" Solution of alkali added to the sap evolved a very fetid odour, but did not appear to exert any particular action on the caoutchouc.

" The sap, left to itself for several days, gradually separated into two parts : the opaque portion contracted upwards, leaving beneath a deep brown, but transparent, solution, evidently containing substances very different in their nature from caoutchouc itself, and which, considering the specific gravity of the sap and of pure caoutchouc (the latter being lighter than water), were probably present in considerable quantity.

" It was found that, by mixing the sap with water, no other change took place than mere dilution. The mixture was uniform, and had all the properties of a weak or thin sap. Heat, evaporation, acids, and alkalies produced the same effects, generally, as before.

" When the diluted sap was suffered to remain at rest, a separation soon took place, similar to that which occurred with the native juice, but to a greater extent; a creamy portion rose to the top, whilst a clear aqueous solution remained beneath. Hence it was found easy to wash the caoutchouc, and remove from it other principles which had been generally involved in it to a greater or smaller extent during its coagulation. For this purpose a portion of the sap was mixed with about four volumes of water, and the mixture put into a funnel, stopped below by a cork : in the course of eighteen or twenty-four hours, when the caoutchouc had risen to the top, and occupied about its original volume, the aperture at the bottom of the funnel was opened, and the solution drawn off; more water was then added to, and mixed with, the caoutchouc, and the operation repeated; and this was done four or five times, until the water came away nearly pure. During the latter washings, the caoutchouc required a longer time to rise to the surface, in consequence of the decreasing specific gravity of the solution in which it was suspended. This was obviated at times, according to the experiments for which the caoutchouc was required, by performing the first washings with solutions of common salt, muriatic acid, &c., and ultimately finishing with pure water.

" In this way the caoutchouc was purified, without any alteration of its original state. It now appeared in its state of mixture with water, perfectly white : portions of it left for a twelvemonth over water underwent no change in that time, except coagulation and a slight film upon the surface ; the rest was as miscible with the water as at first, and, when coagulated, equally elastic. The sap, or the washed caoutchouc, is much more easily preserved in the diluted than in the concentrated state.

" It produced no particular appearance with the solutions of iron or other metals.

" When evaporated, either on paper, or in a capsule, or otherwise, the caoutchouc was left in its elastic state, and perfectly unaltered, except with respect to purity. When put on to absorbent surfaces, as bibulous paper, chalk, or plaster of Paris, the water was rapidly abstracted, and the caoutchouc almost immediately united into a mass, retaining the form of the thing on which it was cast. Mr. Hancock in this way has made beautiful medallions with the sap. Poured on to a filter, the water passes through, and the caoutchouc coagulates.

" When aggregated in any of these ways, the caoutchouc appears at first as a soft white solid, almost like curd, which by pressure exudes much water, contracts, becomes more compact, has acquired elasticity, but is still soft, white, and opaque. The opacity belonging to it is not an essential property of the body, but due to water enclosed

within its mass; further exposure to air allows of
the gradual dissipation of this water, and then the
caoutchouc appears in its pure and dry state, as a
perfectly transparent, colourless, and elastic body,
except it be in thick masses, when a trace of colour
is perceived. The change from first to last is best
seen by pouring enough of the pure mixture into a
Wedgewood or glass basin to form ultimately a
plate of one-tenth or one-twelfth of an inch in
thickness, and leaving it exposed to air at common
temperatures undisturbed.

" No appearance of texture can be observed in the
pure transparent caoutchouc; it resembles exactly
a piece of clear strong jelly. All the phenomena
dependent upon its elasticity, which are known to
belong to common caoutchouc, are well exhibited by
it. When very much extended, it assumes a beautiful pearly or fibrous appearance, probably belonging to the effects which Dr. Brewster has observed
elastic bodies to produce, when in a state of tension, upon light. When it has been extended and
doubled several times, until further extension in
the same direction is difficult, it is found to possess very great strength. Its specific gravity is
0·925, and no reduplication and pressure of it
in a Bramah's press was found permanently to
alter it. It is evidently pervious to water in a
slight degree, or otherwise the interior of a piece of
caoutchouc coagulated from the sap would always
remain opaque. It is equally evident that water
passes but very slowly, from the time it takes to

evaporate that which lies in the middle of a thin cake. It is a non-conductor of electricity.

" The pure caoutchouc has a very adhesive surface, which it retains after many months' exposure to air. Its fresh-cut surfaces pressed together also adhere with a force equal to that of any other part of the piece.

"A strip of it boiled in solution of potash, so strong as to be solid when cold, was not at all affected by it, except that its surface assumed a pearly or tendinous appearance ; no swelling or softening, above what would have been produced by water, occurred.

" The combustibility of caoutchouc is well known. When the pure substance is heated in a tube, it is resolved into substances more or less volatile, with the deposition of only a small trace of charcoal; at a higher temperature it is resolved into charcoal and compounds of carbon and hydrogen ; it yields no ammonia by destructive distillation, nor any compounds of oxygen, and my experiments agree with those of Dr. Ure, in indicating carbon and hydrogen as its only elements. I have not, however, been able to verify his proportions, which are 90 carbon, 9·11 hydrogen, or by theory nearly 3 proportionals of carbon to 2 of hydrogen, and have never obtained quite so much as 7 carbon to 1 of hydrogen by weight. The mean of my experiments gives —

Carbon	-	6·812	} or {	8 proportionals nearly
Hydrogen	-	1·000		7

" No means which have yet been discovered seem competent, when the caoutchouc has once been aggregated, to restore it to its pristine state. Previous to its aggregation it may be either scented or coloured. A solution of camphor in alcohol was added to water, so as to precipitate the camphor in a flocculent state; a little of this was added to the pure caoutchouc in water, well agitated, and then coagulation caused by heat or absorption: the caoutchouc obtained was highly odorous.

" In the trials made to give it colour, the body-colours were found to answer best:—indigo, cinnabar, chrome-yellow, carmine, lake, &c., were rubbed very fine with water, then mixed well with the pure caoutchouc, in a somewhat diluted state, and coagulation induced either upon an absorbent surface, or otherwise. Perfectly coloured specimens were thus obtained.

" The liquid obtained either by letting the sap stand for some time, or by the first and second washing, was of a brown colour, bitter, acid to litmus, in consequence of the presence of acetic acid, due apparently to spontaneous changes in the substances present. It was difficult to filter. Being boiled, acid vapours rose, a precipitate fell to the bottom, and now the solution became clear, either by standing or filtration, and could be separated from the solid matter.

" The precipitate or substance thus obtained was dark brown, glossy, and brittle, much heavier than

water, not soluble in alcohol, ether, water, essential
or fixed oils. Weak solution of alkali dissolves it,
forming a deep brown solution, precipitable by
dilute muriatic acid. It burns upon platina-foil,
like animal matter, with flame, leaving a bulky
charcoal. When heated on a tube, it chars, yielding much ammonia. It resembles albumen more
than any other substance, and is the source of the
nitrogen or ammonia obtained by the distillation of
common caoutchouc.

"The brown aqueous solution becomes frothy
on agitation; alkali rendered it of a deep yellow
colour, and produced a putrescent odour, similar
to that evolved by alkali, or quick lime, from white
of egg, or blood. It was remarkably distinguished
by the deep green colour it produced with per-salts
of iron, especially when a little alkali was present,
and the dense yellow precipitates it formed with
muriate of zinc and nitrate of lead; indeed, precipitates were produced in solutions of most of the
metals by it. The colour produced with iron does
not seem to be a precipitate.

"With the hope of obtaining something peculiar
from this solution, a quantity of it was precipitated
by nitrate of lead; a colourless solution and a
yellowish green precipitate were obtained. The
latter, being well washed, was next diffused through
water, and sulphuretted hydrogen passed through
it; by filtration a deep brown solid was obtained,
and a yellowish solution. The precipitate when
washed and dried was brittle and hard; on platina-

foil it at first burnt with flame, swelling much, and
giving out odour of ammonia like animal matter;
after that, sulphurous acid burnt off, and ultimately lead and oxide of lead remained; hence it
was a combination of sulphuret of lead and a
highly azoted substance. Heated in a tube, it
gave out much ammonia; digested in alcohol,
scarcely a trace of matter was removed.

"The sulphuretted hydrogen solution being boiled
and evaporated, left a yellow varnish-like substance
not deliquescent, soluble in water, acid to taste and
to litmus, the acid not being sulphuric; it rendered
per-sulphate of iron green, precipitated nitrate of
lead, and gave no ammonia by heat.

"The concentrated solution acted upon by
alcohol had an insoluble matter thrown down,
which, being separated and well washed with alcohol, was afterwards treated with water; a deep
brown aqueous solution was obtained, and a
small insoluble portion left; this was almost black
when dried, tasteless, brittle, burning with difficulty, and when heated in a tube giving much
ammonia.

"The solution was almost tasteless, and when
dried left a green, shining, brittle substance, re-
soluble in water, and of course precipitable by alcohol. It colours solution of per-sulphate of iron
green; but if its strong aqueous solution be treated
with muriatic acid, a reddish brown precipitate is
formed, which, when separated, dissolves in water,
does not colour per-salts of iron, and when evapo-

rated yields a pulverulent substance, burning, but not with facility, and producing a little ammonia when heated in a tube.

" The alcoholic solution from which these matters had been separated contained the particular principle which colours per-salts of iron green. When evaporated, it left a brown, brittle, transparent, substance, becoming soft by exposure to moist air. It is very bitter, soluble in water, &c., slightly acid. When heated on platina-foil it does not burn easily, but runs out into a bulky charcoal much like animal matter; at the same time it does not yield ammonia when heated in a tube *per se*, though the smell is very animal.

" Ether warmed with it dissolved a small portion of matter, and the solution, upon evaporation, left globules, which in all their characters corresponded with wax : its quantity was but small.

" Nine hundred and eighty-one grains of the original sap were washed in water several times. The washed caoutchouc, being coagulated by heat and perfectly dried, weighed 311 grains. The aqueous solutions, upon being boiled, yielded sufficient of the heavy precipitate to equal, when dried, 18·6 grains. The clear solution was now evaporated to dryness, and digested in alcohol; 28·5 grains of insoluble matter were left, and the solution, upon evaporation, afforded 70 grains of dry matter. Hence the following are the contents of nearly 1000 parts of the original sap.

Caoutchouc - - - 317·0
Albuminous precipitate - - 19·0
Peculiar bitter colouring matter, a ⎤
   highly azotated substance - ⎬ 71·3
Wax - - - - ⎦
Substance soluble in water, not al-
   cohol - - - - 29·0
Water, acid, &c. - - 563·7
                      ——
                   1000·

" Thinking it probable that whilst in its natural state of division the caoutchouc would combine more intimately or readily with fixed and volatile oils than when aggregated, as it generally is in commerce, an experiment or two were made in consequence. A portion of well-washed milky caoutchouc being added to olive-oil, and the two beaten well together, a singularly adhesive stringy substance was produced, which, holding the water diffused through it, assumed a very pearly aspect, stiffened, and was almost solid; upon being heated so as to drive off the water, it became oily, fluid, and clear, and was then a solution of caoutchouc in the fixed oil. On adding water and stirring considerably, it again became adhesive as before. Thus introduced, caoutchouc would probably be a useful element in varnishes.

" Oil of turpentine being added to a mixture of one volume of sap and one volume of water, and well agitated with it, was found to be only imperfectly miscible: after standing twenty-four hours, three portions were formed; the lower, the usual aqueous solution; the upper, oil of turpentine,

holding little caoutchouc in solution; the intervening part, a clot or tenacious mass, soft and adhesive,
like bird-lime, consisting of caoutchouc, with some
oil of turpentine. It was very difficult to dry, and
always remained adhesive at the surface; but experiments of this kind were not pursued, for want,
at that time, of further quantities of the original
sap.

" Such is a general view of the nature of the sap
from which the substance is obtained, and of the
substance itself. I have not endeavoured to give
an accurate account of the properties or quantities
of the other substances present, because there is
reason to believe that both vary in different specimens, probably according to the age of the tree,
the time of the year, or the manner in which the
sap is drawn; nor have I dwelt upon the inaccu-
racies of former accounts, inasmuch as they are
evidently referable to the impurity of the substance examined."

I will now return to the double textures. I
have said little yet of a somewhat important part
of these manufactures known as pneumatic articles.
I think I began with making beds capable of being
inflated, and sufficiently strong and air-tight to
sustain the necessary weight: or, the first may
have been a pillow, cushion, or life preserver; for
all these were made about the same time, and on
similar principles. I will first take a bed: this was
made in only one compartment, so that, when
inflated, it assumed a pin-cushion shape, and its

rotundity was such that to place yourself upon it
and remain there was impossible; try as you
might to balance yourself, in a moment you lost
your equilibrium and came rolling on to the floor;
each bystander thought he could do it, but the air-
bed set him tumbling about, and all at length
acknowledged a defeat, and declared that air-beds
" would never answer." A portion of the air was
let out, but the same kind of objection remained;
and although this was repeated until nearly the
whole of the air was exhausted, this principle of
construction was still evidently defective. The
same inconvenience was found to attend an air-
cushion : — you could not sit still a minute ; make
but the least alteration in your position, and you
commenced a rolling motion, to which there seemed
to be no end as long as you sat there. This was
a defeat, and, if I remember right, the first
attempt to remedy it was a really good one, as it
fully answered the purpose, and possessed many
advantages, although it has not been much in use.
It consisted in preparing a case of ordinary bed-
ticking divided into seven or eight compartments ;
an air-proof cylindrical chamber of a proper length
and diameter was made for each of these compartments, and inflated to any desired degree. This
was an obvious improvement, and the air-bed thus
constructed, whilst it yielded sufficiently to the form
of the body, supplied at the same time a more
elastic resting-place than ever the human form had
before reclined upon. There was another advan-

tage in this mode: if any accident, allowing the air
to escape, happened to one or two of the cylinders,
they could be placed under the feet, where they
would be subject to less pressure: a defective
cylinder could be sent for repair, and the bed still
be tolerably effective. This old mode of construction has since been applied to nautical contrivances,
and is apparently supposed to be a new idea!

Cushions were soon made on the same principle
with equal success. This mode, however, did not
last long; it was said that one of the main advantages derived from the employment of air for these
purposes was lost; in other words, they were not
so portable in this form as they might be; the
outer case was simply an encumbrance. This
objection gave rise to another contrivance: — the
framework (if I may so term it) intended for the
interior was made of the form and size of the
intended cushion, which was divided into small
compartments, by sewing in partitions that were to
limit the thickness of the cushion. These partitions
were all laid down flat, and the whole stretched by
tacks upon a board, and then paid over with a
coating or two of solution; when dry, it was turned
over and treated in the same way on the other
side: when again dry, this frame of the cushion
was laid upon a piece of jeanet, previously rendered air-proof, which projected all round half an
inch beyond the frame; the cushion was turned
over, and a similar piece of air-proof cloth laid on
that side: the whole was then well rubbed down,

AIRPROOF GOODS.

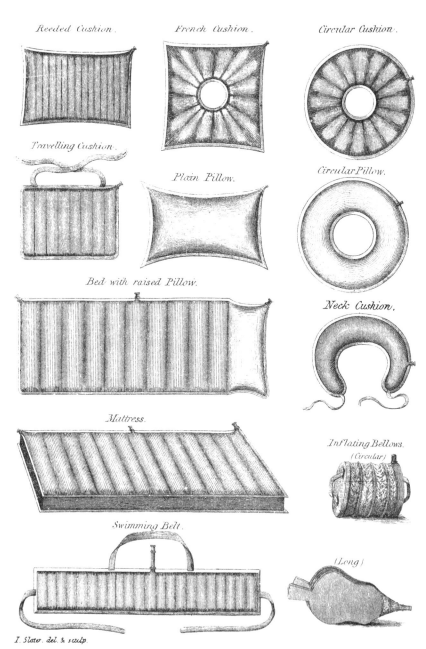

Reeded Cushion.

French Cushion.

Circular Cushion.

Travelling Cushion.

Plain Pillow.

Circular Pillow.

Bed with raised Pillow.

Neck Cushion.

Mattress.

Inflating Bellows.
(Circular)

Swimming Belt.

(Long)

I. Slater. del. & sculp.

and the projecting edges of the air-proof cloth carefully united. This cushion, when inflated, formed
nearly a flat surface on each side, excepting that
where the partitions occurred there was a slight
indentation, bringing the surface into a kind of
reeded form; and hence such cushions were ever
after called "reeded cushions." Nothing apparently
could be better than this, and these cushions in
general seemed to give satisfaction. Other forms
of surface for cushions were adopted, but this form
and one other, which was depressed in the centre,
with radiating partitions, took the lead. There
was, however, yet one great improvement in the
economy of their manufacture, which consisted in
dispensing entirely with the sewing, and employing
solution as a cement universally in its stead.

Pillows were generally preferred with partitions;
but a kind of collar-cushion for the neck had no
partition. Life-preservers were at first made of a
cylindrical form, but afterwards flat in the reeded
form; as were beds also.

Pillows, cushions, beds, and life-preservers, were
externally made of cambric or jeanet in general;
but pillows and cushions had sometimes an external
covering of silk, or kid and morocco leather, some
of which latter were used by George IV. in his
last illness. At first the aperture for admitting the
air was simply a stop-cock; but, after a variety of
methods had been tried to supersede it, the screw-
valve at present in use was adopted, and is universally preferred.

E 2

For an easier method of inflating beds and other articles of large capacity, a kind of cylindrical bellows was made of air-proof cloth.

Late in the year 1825 it was proposed that Messrs. Macintosh and Co. and myself should come to some arrangement by which articles made under Mr. Macintosh's patent should emanate from the establishment of the firm; and, during this correspondence, I stated in some detail to Messrs. Macintosh and Co. the large extent to which I contemplated carrying out my views in regard to the applications of rubber, and the patents I had taken out to secure them, and that it would be necessary for me to engage with capitalists ready and willing to co-operate with me. This correspondence resulted in an arrangement in February, 1826, by which I engaged to manufacture for Messrs. Macintosh and Co. the articles covered by Mr. Macintosh's patent, providing that a partnership should be avoided (to which I ever had a dislike), that my name should be stamped on all the goods I made for them, and that no other goods but those made by me should be sold within the limits of the bills of mortality. This arrangement did not interfere with my business in other respects; for the present our relations extended no further.

I should mention here that neither the firm nor myself ever intended to have any retail shops; and we desired by all means to avoid the making of garments, and wished to sell our goods only in a warehouse; but we were compelled to do all three, or lose our business. Some — most — of the tailors set their faces against the use of our material; others made it up so badly that the garments were not waterproof: at every seam the cloth, being necessarily punctured by the needle, allowed the water to pass. Our advice was to make no close garments, and as few seams as possible; and to enable them to do this, we furnished the cloths wide enough to make the length of cloaks and capes: but they persisted in making garments to sit close, and were greatly offended when told that they could not sew a water-tight seam, and that it was necessary to send their garments to us to have the seams lined to make them proof. Some of them persisted, and actually made a double row of stitches to make sure work of it! We tired of all this, and opened retail shops, and employed our own tailors, and proofed our seams; and, even then, so accustomed were these men to pin their work, that we very frequently found pin and needle holes in the body of the cloth. This business of securing stitched seams against the entrance of rain proved extremely difficult for a long time, even to ourselves: constant complaints were made of the water coming through, even in those cases where the seams had been treated by our most earnest and careful workpeople. It must be understood that in double textures the seam-proofing was done inside; we at length, however, discovered, after repeated observation, research, and much attention to the subject, that although the direct

E 3

entrance of the water was stopped by the means adopted, yet that the thread of the tailor, taking up the moisture, became the medium of conveyance to the threads of the inner cloth of the double texture, and that the water so introduced was carried by capillary attraction along the warp or weft of the cloth, and so spread over an extent of surface depending upon the length of time for which the garment was exposed to the rain, and the absorbing capacity of the tissue for taking it up, which not unfrequently would allow of the water spreading over nearly the whole surface: no doubt in many cases the insensible perspiration of the wearer met the rain half way, and increased the evil. Complaints arising from this source long annoyed us, and exposed us to no end of abuse, whilst we were using our utmost efforts to furnish to the public garments which should be a certain protection from a wet skin. I mention this as one, only one, of countless difficulties that for a long time attended us in the first years of our progress ; and, as we proceeded, other difficulties, losses, and vexations followed us up, of which I could give a long list, but it would only tire the reader.

We were most fortunate in obtaining a superintendent in London of first rate ability, not only as a man of business and integrity, but uniting firmness of character with conciliatory manners ; and knowing well the aim and intentions of his employers, he conducted our affairs with the greatest prudence and discretion, and indeed the nature of the new kind of business he had to conduct required the ability of such a man, and I feel pleasure whilst I write in bearing my testimony to the value of the long and tried services of Mr. S. Matthews·

I may here mention that a waterproof military cloak of blue cloth, lined with crimson silk, had been made for the Duke of York, and the officers of the Guards began to wear light drab cambric capes on their way to field exercise, and other young men as usual following their example, our material (especially of this drab colour) began to take with the public generally, and more and more as the value of it, and its really waterproof quality, became known.

Before I close this part of my narrative I may mention that which occasioned us much trouble and annoyance,—the persistence of the public for a long time in having garments made to fit too close, which brought the material into some degree of disrepute from the want of free escape of insensible perspiration when taking active exercise. Time only demonstrated the absolute necessity for wearing them large and loose, as now universally practised.

As we progressed with our manufactures, great improvements were introduced into air-proof articles. These were first appreciated by invalids, not only as ordinary beds, pillows and cushions, but in a variety of forms to be used by bed-ridden patients, and in the hospitals, and in lieu of splints, pads, bandages, &c. by surgeons, who also began

to recommend their use in carriages; so that at
length this branch of our business became a very
important one, and served at the same time to
bring the knowledge of the peculiar qualities of
our manufactures before the public.

Although so far successful in the main, yet
further experience brought new difficulties and
vexations to light: — we frequently had garments
returned defective in the waterproof qualities of
the cloth itself : the cause of these defects was not
easily discoverable ; great pains were taken to trace
back the defective articles to the pieces from which
they had been cut, and then to the mill-books, to
ascertain if possible if any deviation from the usual
course could be discovered. At length an accidental circumstance enabled us to trace the source
of one cause of evil, which I will explain. In
order to expedite the work, it had become the
practice to unite a number of pieces of cloth
together at the ends, so as to form them into one
length; and in this way the whole passed through
the machine together. It so happened that a case
occurred where one or more of these pieces turned
out defective by decomposition, whilst the rest remained sound ; and it was ultimately proved that
the defective pieces, being worsted goods, had not
been properly scoured, but retained in the body of
the cloth remains of greasy matter.

We had become well aware that grease acted injuriously on rubber, particularly on thin films : we
had therefore given special directions to our manufacturers, and repeatedly cautioned them to be most

particular in attending to this matter, foreseeing
the mischief any carelessness would occasion. On
examining our stock we found the rubber on 900*l.*
worth of goods in a state of incipient decomposition. This serious matter was referred to arbitra-
tion, and we recovered that sum, and destroyed
most of the goods. We had many cases of this
sort, and the reputation of our goods suffered from
this cause, besides occasioning us great annoyance,
trouble, and expense; as we met every reasonable
case by exchanging the article. Another cause of
decomposition was the use of chemical processes in
the preparation of some of the colouring matters
used by the dyers of our cotton goods. These had
to be searched out until we ascertained what
colours we could safely use. The goods so rendered defective were destroyed.

The injurious effect of the sun's rays upon thin
films of rubber we discovered and provided against
before much damage accrued. All these things
are now cheaply known to those who have followed
us by men leaving our employ, and the specifications of our patents; but they had all to be undergone in our early progress at an enormous cost, as
well as of trouble and vexation: and none but
those who have passed through the ordeal can conceive the mortification experienced during those
years; for whilst we were, regardless of cost,
making every possible exertion to introduce an
article confessedly much wanted and long sought
for, and operating upon a new material, hitherto

comparatively little known in the arts, and surrounded with unforeseen difficulties, an impatient
public gave us little credit for our exertions, and
persisted in attributing that which was our mis-
fortune in any occasional failure to any cause but
the true one; being nothing less forsooth than an
imposition upon their complacent and confiding
good nature!

In the year 1826 bags began to be made for
containing gas for temporary illuminating purposes,
and we have continued making them up to this
day, for various experimental purposes, and also
for the popular exhibition of the oxyhydrogen
microscope; but the first was made in May 1826,
at the suggestion and for the use of the late
Lieutenant Drummond, during his trigonometrical
survey.

These were made of very strong materials, not
only to enable them to sustain the internal pressure of the gas, but the rough usage they were
likely to be exposed to in such a service. The air-
proof lining was of thin cut sheet-rubber, and the
exterior of fustian. I had the curiosity at the
time to make a bag of this material, which I filled
with water and sealed hermetically. I did this for
the purpose of discovering whether rubber is or is
not absolutely impervious to water. I suspected it
was not. This bag is now before me, and I will
copy from the record written upon it of its original
weight when filled, and the periodical decrease of
the water contained.

				lb.	oz.	dr.
Oct. 21st, 1826	-	-	-	1	1	4
Oct. 25th, 1827	-	-	-	1	1	2
Oct. 2nd, 1835	-	-	-	1	0	0
Nov. 1844	-	-	-	0	14	12
Oct. 1849	-	-	-	0	13	4
Feb. 1851	-	-	-	0	7	8
May, 1854	-	-	-	0	3	14

I have just now, 1856, cut it open; it is quite
dry, and weighs three ounces twelve drachms,
proving that rubber is not absolutely impermeable
to water, but admits of a slow and gradual absorption of moisture through its substance; and in this
case the whole of the contents of the bag escaped, or
rather more than twelve ounces, in the long course
of twenty-five years! Bags made of air-proof cloth,
that is, with only a thin coating of rubber, soon
evaporated sufficiently to moisten the cloth, when
the bags were piled upon each other, and produced
mildew. This slow evaporation does not interfere
with its efficiency for ordinary purposes.

Captain Parry, in the Narrative of his Attempt to
reach the North Pole, in His Majesty's ship " Hecla,"
in the year 1827, thus speaks of this invention, p.
72 : —

"Just before halting at 6. A.M. on the 5th July,
1827, the ice at the margin of the floe broke while
the men were handing the provisions out of the
boats ; and we narrowly escaped the loss of a bag of
cocoa, which fell overboard, but fortunately rested
on a tongue. This bag, being made of Macintosh's
waterproof canvas, did not suffer the slightest

injury. Of this invaluable manufacture, which consists, I believe, in applying a solution of elastic gum or caoutchouc between two parts of canvas, it is impossible to speak too highly. I know of no material which with an equal weight is equally durable and water-tight — in the latter quality, indeed, it is altogether perfect, so long as the material lasts."— *Narrative of an Attempt to reach the North Pole in Boats attached to His Majesty's Ship " Hecla" in the year* 1827, by Captain Parry, R.N. F.R.S. London: 1828.

About the year 1825 tubes made of rubber came into some demand for surgical and other purposes. About this time a popular toy was introduced: it was made by inserting a condensing syringe into the mouth of a bottle of ordinary rubber; the bottle was kept ordinarily warm whilst the injection of the air proceeded, until at length it became so much extended as to form a tolerably strong semi-transparent sphere, and made a good nursery play-ball for children. Some were made of a large size to be attached to the blow-pipe of the chemists: they were in fact made of all sizes, and afterwards, when the supply of bottles fell short, means were found to use flat pieces for the purpose. They were sometimes ornamented by painting and by covering them with fanciful network. This toy had a prodigious run for some years, and the trade in them has been several times revived: they still continue to be made.

The cut sheet rubber began about this time,

1825, to be used for a number of surgical purposes, as before mentioned. All these kinds of articles were taken up by my late brother, John Hancock, and manufactured by him from materials supplied by me. Those for surgical purposes need not be enumerated here, although some of these applications were important and still continue to be made. He also devoted a great deal of his attention to the manufacture of tubing, made sometimes solely of sheet-rubber, and also by uniting plies of cloth. This gradually led to the introduction of rubber hose-pipe, which met with a vast amount of opposition from the leather-hose makers, and amongst brewers and distillers,—particularly the men, who were in league with the old makers. It appears that the leather-hose used in breweries could not be made so perfect as to prevent a great loss of liquor. It was stated to one of the great firms (Messrs. Barclay and Co.) that rubber-hose would not admit of any escape ; and one of the firm insisted upon its being tried : it was found effectual, but imparted a bad flavour to the liquor. This was for a time a great obstruction to its use ; but it was found that, by allowing waste liquor to run through the hose for a while, it became sweetened, and ultimately (if I mistake not) about 1800 feet was constantly in use in that brewery alone. Others followed, and rubber hose and tubing became a staple manufacture, and continues so to this day. The hose-pipe was composed of two, three, or more plies of cloth, coated on both sides with solution, and then rolled up on mandrels. They were generally lined with sheet-rubber; suction-hose had spiral wire inside to prevent it from collapsing.

Shoes began to be made of the cut sheet-rubber, I think as early as 1825 or 1826, but not in considerable quantities till 1828 and 1830. They were made by folding the sheet, cut to the form of an " upper," over to the bottom of a wooden last, and held there by a few stitches across from edge to edge ; and then the sole, cut to its proper size and form, was cemented on with solution. These shoes were generally lined with cloth and leather. The soles were sometimes made of ordinary sole-leather and cemented on with solution. Several parties soon began to make these shoes, but in a rather clumsy manner: there were exceptions, however, and some of very tolerable form. Mr. Sparkes Hall soon took the lead, and maintained it. He has told me that he began in 1830, and made twelve pairs a day with his own hands. All these parties took their cut sheet-rubber of me: indeed there was yet no other manufacturer of this article, the sale of which was become rather large ; and for this and the other branches of my rubber business I now constantly employed four horses.

About this time I supplied the Board of Ordnance with some waterproof calico, which I understood was applied to large and small cartridges, to prove whether the powder would be kept dry by using this material. I expected some good orders from these experiments, but I do not remember ever hearing anything more about it. The only articles with which we at first supplied the Government were " saddle water-decks," for covering the saddle, &c., when the soldier was dismounted; and we have supplied these articles at various periods ever since. We also used air-proof cloth about this time for making the well-known diving dresses, and sheet-rubber tubings for keeping up the supply of air to the diver, and to enable him to communicate with the people in the vessels above.

The manufacture of rubber goods had not yet reached the Continent, and in 1828 proposals were made to me by parties who were desirous of establishing it in Paris. Terms were soon agreed on, and I immediately put such machinery in hand as the nature of the arrangements required. I was to supply the solution from England, so that I did not impart to them my secret mode of masticating the rubber, nor of making the solution. I engaged a sufficient number of men, and instructed them at my own works in the modes of applying the solution, and doubling the cloths, and also in the manufacture of air-proof beds, cushions, &c.

In the summer of 1828 I took out to Paris Mr. Christopher Nickells and Mr. Edward Woodcock, Jun., and several other subordinate work-people, and the machinery I had prepared, but found it immensely difficult to pass the machinery through the Custom House; and I was obliged to wait at Calais whilst communications passed between the officers there and the superior authorities in Paris. At length, after unpacking the whole, and explaining the parts to an engineer who, I believe, understood nothing of my explanations, the whole was allowed to pass. We had similar difficulty with the solution: they opened the casks, put in a stick, and stirred it to the bottom; and, on withdrawing it covered with the solution, they smelt its disagreeable perfume, turned up their noses, exclaimed "Chimie," and let it pass.

I met with a most cordial reception at Paris, as well as my men; and after fixing the machinery, and seeing all in operation, and the men well to their work, I came home, having spent, I think, about three weeks in Paris, participating in the hospitality of my new friends, Messrs. Rattier and Guibal, whose attentions and politeness I shall never forget.

The manufacture commenced at St. Denis, near Paris, in the beginning of July of the same year, and the first piece of cloth was waterproofed by Mr. E. Woodcock. The shop for the sale of goods was opened in the Rue de Fosses Montmartre in October. Mr. Woodcock (except during a short interval) continued in the employ of the firm until its dissolution, and remains the superintendent of the successors of the business to the present day; being an ingenious man and skilful mechanician, he has contributed perhaps more than any other person to the introduction of useful applications of the rubber manufactures in that country.

Mr. Nickells remained also in the service of Messrs. Rattier and Guibal some years, and has since, with some others in this country, pursued the elastic web and other manufactures with great success.

I continued to supply the solution for some considerable time, and at length furnished the means for making the solution in France. The difficulty of procuring very pure oil of turpentine caused some loss and disappointment; but this was ultimately overcome, and very excellent goods were manufactured. The sale of waterproof cloths and pneumatic articles did not, however, make much progress at first; but the adaptations of the material to some surgical purposes were eminently successful.

In 1830 I took out another patent for the application of the pure liquid rubber as drawn from the trees to a number of useful purposes, as described in the specification of this date in the Appendix. This sepcification was settled by Mr. Serjt. Bompas. I was induced to take this patent from the persuasion that this article, having once been brought over in the liquid state, could be brought again, if proper steps were taken to procure it. On the recommendation of a friend, I employed a person in Tampico to conduct this business, and sent him instructions for collecting it according to the best information I had been able to obtain. A large quantity was collected and sent over in good sound barrels, well stopped; but, on

opening, I found in the majority of them a solid
mass of good rubber and a brown fluid. On inspection it was evident that the solid part of the
rubber had separated from a dark-coloured watery
fluid, and had taken the form of the end of the
barrel, where it was found deposited. In some of
the barrels, however, the separation had only partially taken place, and I obtained some rubber in
the creamy state. Another lot arrived much in the
same condition; and such were the expenses in-
curred and the loss sustained, that I gave up the
attempt, and all my patents for the application of
the liquid expired before I could obtain it in any
sufficient quantity. Nevertheless, I am persuaded
that its importation is practicable, as I have in my
possession a small quantity which I treated in a
particular way many years ago: and it remains a
fluid to this day. I may also mention that several
barrels were afterwards sent to this country, which
had undergone some treatment in the country
where it had been produced, by which it was preserved in a kind of semi-fluid state, and was in the
market here, and purchased by several persons.
Samples have recently been brought here and
proved satisfactory: these last had been treated
with an admixture of ammonia.

Although rubber in this state would be very useful, and many things could be done with it which
are hardly practicable with the solutions; yet the
loss of weight by evaporation being nearly two-
thirds of the whole, the expense of vessels, and the

freight of so much worthless matter, will probably prevent its ever being used extensively. Before the difficulty of dissolving ordinary rubber was overcome, it was thought that the liquid, if it could be obtained, would be invaluable; but now, all things considered, the dry material, for nearly all the purposes of manufacture, is the cheapest and most easily applied; although, to persons unacquainted with practical details, this may appear enigmatical. I have made very sharp and clean casts with this liquid, and as it is susceptible of tinting with delicate colours, it might, for ornamental purposes, be rendered very beautiful.

I have before observed that my mode of introducing rubber elastics into articles of dress and wearing apparel became ultimately superseded by an improved mode of applying it. I have understood, that a German whose name I am not acquainted with, conceived the idea of introducing a thread of rubber into a woven web or fabric, so as to form the warp, and, by keeping it confined in an extended state during the operation of weaving, and then releasing it, the fabric would be gathered up and elasticated.

It appears that he was at a loss how to cut the thread, which proved a difficulty which he could not overcome; and to obtain assistance, he went to Paris, and, I believe, communicated to my friends there his invention, and his difficulty. Experiments were commenced at their works, and the person I took out with me and left to superintend the waterproof manufacture succeeded in producing a thread of rubber. I think, if I am rightly informed, the little sample originally produced by the German had been covered with a thread of cotton or silk: at all events it was thought necessary then, and for some time after, that the thread of rubber should be covered in some way, and the braiding machine was put in requisition for the purpose.

Soon after this elastic web appeared in this country, I had some made, the cutting of the thread not being very difficult to me. A very pretty article was soon produced: the only obstacle to success lay in the weaving, an art with which I had very little acquaintance; and when a practical weaver was found, who could weave it well, she was greatly perplexed to keep all the threads, or elastic warps, to an equal tension. The consequence was that the web, when taken out of the loom, ran up into a crooked serpentine form, caused by the threads which were more extended than the rest during the weaving contracting more when set at liberty. All this was matter of detail, and a more fitting business for a ribbon or tape manufacturer. I accordingly entered into a mutually beneficial arrangement with an eminent house of Manchester in that line, supplying them with the rubber thread (then called gut), which they soon manufactured into beautiful webs of different widths and different degrees of elasticity and strength. This arrangement subsisted for some time; but having now a great deal to attend to, I agreed upon terms to relinquish the whole business in their favour, furnishing them with my modes of preparing and cutting the thread, and two of my men to carry it on. Their practical knowledge in braiding and weaving enabled them to perfect the manufacture.

In order to show, as I before observed, the step-by-step progress of new manufactures, I will devote a few lines to the subject of cutting rubber-thread, which may produce a little feeling of complacency in those who now get this material furnished to them in so perfect a condition without trouble.

My manufactured sheet-rubber was too inferior in elastic power, when cut into such minute dimensions, to produce a good result; and although web has since been made of masticated rubber, it did not add to the reputation of the article in its elastic qualities, whatever it may have done in regard to an even surface and a neat appearance.

I began, therefore, to cut the thread from the rubber as imported, choosing the best quality and the largest and thickest of the bottle kind. After trying various modes, the following was practised for some time. I must premise that these bottles were very irregular in form, and of different sizes and thicknesses. After softening them in hot water, they were cut through the middle lengthwise, and placed between plates, and submitted to pressure to flatten them, and remained

until cold.   They were then cemented to a board,
and ready for the cutting machine, which was
a kind of lathe carrying a circular knife, the
edge of which just came in contact with the
surface of the board, on which the flattened halves
of a number of bottles were cemented: a slide
movement carried the board and rubber past the
knife, which, having a suitable motion communicated
to it and water dropping on it, made a clean cut
through the rubber, near to its edge: a screw motion
now pushed the board and rubber as far beyond
the knife as would produce the thread of the intended thickness to be cut.   The rubber was then
pushed past the knife as before, and so on, until
the whole breadth of the rubber was filled with
these cuts (which were in general about one-sixteenth of an inch apart).   The next operation was
to put the board and rubber into a machine exactly
similar to that for cutting sheets ; after a few cuts
to level the surface, the screws were made to raise
the rubber, say one-sixteenth of an inch above the
steel-plates, and then passing the knife through the
rubber as in cutting sheets.   The threads then came
off in a square form, of the intended size.   These
threads averaged about five inches in length, and
when hot might be extended to about a yard.   Now
these threads, being intended for warps, were required of great length, and, to obtain them, the
short lengths had to be united ; this operation was
done very neatly and quickly by girls ; with a
pair of sharp scissors they cut each end of the

thread wedge-shaped, and, when warm, they were brought into juxtaposition splice-fashion; and then, giving them a rub, they were perfectly joined to any length.

A great improvement was afterwards made by putting the rubber bottle on a mandrel in a lathe, and by means of a circular knife and a screw and slide motion a tape-like long slip was obtained; and this was again cut into square threads in another machine, and then united as before, but with fewer joinings, and consequently less labour and cost. Some time before I gave up this manufacture I had a cylinder made of masticated rubber of such a size as would be most convenient for this purpose, and sent it to Para as a pattern for the natives; and great numbers of such cylinders were soon after in the market, and were well made, and of the best quality of rubber, as desired, and called tubes. Such cylinders are still imported. Before I quit this web-business, I will mention a peculiar property in rubber, which was taken advantage of, and was of great importance at that time. The threads, during the weaving, had to be kept equally stretched, which was a difficult matter. This was obviated thus: — the thread was immersed in hot water and stretched out to its utmost tension, and kept in that position on frames. After standing by for some days, it became " set " (that is, it remained so extended by the action of cold) and was transferred to the loom. When the web was taken out, a hot iron was passed over it, when the restored resi-

liency of the threads contracted the web to its proper
length.

In 1830 Messrs. Macintosh and Co. made trial
of my solution at Manchester, which resulted in a
proposal to me to manufacture the solution there:
the terms proposed were liberal; and, as we had
now become well acquainted with each other, and
had always gone on cordially together, the arrangement was soon made—but still avoiding a partnership. It was resolved for the present not to
carry on the masticating there, and for some time
the masticated rubber for making the solution was
furnished by me. In pursuance of this arrangement
I now visited Manchester for the first time; and
after surveying the works there, and suggesting
the necessary alterations, I returned, and put the
machinery in hand for making and straining the
solution, and for spreading it on the cloth; and, as
it was necessary still to continue operations in
London, took on fresh hands, who were practised
for some time at my works in making and straining
the solution, and in the use of the machines employed in spreading it on the cloths and doubling
them, in rendering the seams waterproof, and
making boots, diving dresses, and other similar
articles, and air-beds, cushions, pillows, life-jackets,
life-preservers, hot-water bags, &c. As soon as
the works were ready at Manchester, I accompanied the men thither, and soon brought the
whole into successful operation. I met with a
great improvement here in drying with despatch

MECHANICAL PURPOSES.

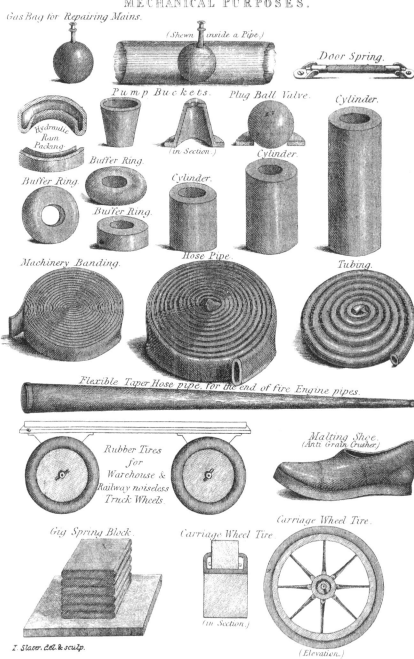

Gas Bag for Repairing Mains.

(Shewn inside a Pipe.)

Door Spring.

Pump Buckets.

Plug Ball Valve.

Cylinder.

Hydraulic Ram Packing.

Buffer Ring.

Cylinder.

Buffer Ring.

Cylinder.

Buffer Ring.

Buffer Ring.

(in Section.)

Machinery Banding.

Hose Pipe.

Tubing.

Flexible Taper Hose pipe, for the end of fire Engine pipes.

Rubber Tires for Warehouse & Railway noiseless Truck Wheels.

Malting Shoe. (Anti Grain Crusher.)

Gig Spring Block.

Carriage Wheel Tire.

Carriage Wheel Tire.

(in Section.)

(Elevation.)

I. Slater. del. & sculp.

the solution after it had been spread on the cloth,
by passing it over large cylinders heated by steam.
Mr. Macintosh had an apparatus applied to these
machines with the view of condensing the vapour
and recovering the naphtha, but found it could
not be done with any useful or economical result.

Although I had so far associated myself in
business with Messrs. Macintosh and Co. in carrying out Mr. Macintosh's patent, I still for some
years continued to conduct my own business quite
distinct from theirs.

In the year 1833, my late brother John Hancock, who had established the hose business, left
London, and sold his interest in the business to
Messrs. Macintosh & Co., and it was then carried
on chiefly at my works, where I continued to make
nearly all the air-proof goods and other similar
things sold in London, as well as water-proofing
the seams of garments, &c.

On the 11th of April, 1834, as some of my men
were working by candle-light, they by some means
set fire to a whole piece of cloth they were
proofing; the flames communicated immediately
with other pieces and with a quantity of solution
they were using, bringing the whole room suddenly
into a blaze. The men, whilst making their en-
deavours to arrest the mischief, were enveloped in
a thick smoke, and, being unable to find the stairs,
were nearly suffocated, and one of them was much
burnt. The fire-engines were soon on the spot; but,
from the combustible nature of the stock, and a

considerable part of the buildings being constructed of weatherboard and plaster, nearly the whole was speedily reduced to ashes. The premises and stock being insured, my loss was not very great. In laying out my plans for the new works, I took care to have the buildings detached, with a considerable space between each of the three, and without windows or doors on opposite sides. These precautions and some others enabled me to get them insured again.

In the new buildings no provision was made for manufacturing either rubber or solution, which was now supplied from Manchester; this alteration afforded me much greater space for spreading machines for water-proofing and air-proofing cloth, and for the manufacture of air-proof articles, proofing the seams of garments, and a variety of other things, the whole business having now greatly increased in magnitude.

I should mention that I had now carried on the manufacture of rubber shoes to some considerable extent; they were lined withcloth, velvet, leather, &c.; some were made with leather soles, and others entirely of rubber. Several things, however, made me indifferent about shoes: first of all, it was not a business to my taste; then shoemakers, who would have been thought the most likely persons to have made them well, were in practice amongst the worst; they could not be reconciled to abandon their awl and waxed thread, although they were not called upon to give up their "last." I must confess also that others made them neater than we did; and as all the makers took their material from us, I relinquished this trade early in the year 1834 without much reluctance.

As it now became necessary to employ steam power, it was determined that such parts of the business as required the employment of it should be done at Manchester. Accordingly, I was now busily engaged in preparing the masticating and other machinery for the works there. The demand for cut-sheets and stationers' rubber was constantly increasing, and required of itself a large amount of power to drive the capacious masticators and the larger rollers now found necessary for this branch of manufacture; and as the waterproof business also kept steadily advancing, we required a large supply of masticated rubber to be used in our own works, for making solution, besides its application to other purposes, such as solid rubber-tubing, the lining of hose-pipe, &c. Sheet rubber was also brought into requisition for lining hot-water bags, which required an extra thickness of rubber. These bags were made of all sizes, and at the breaking out of the cholera and during its pre-valence were in great esteem, and have ever since been more or less in use, and sometimes large enough for beds; they were, and are, used also for cold water by the doctors in particular cases.

Although there was already one steam-engine at the works, another was speedily laid down, to-gether with a train of heavy shafting and gearing to drive the new masticators and rollers; and steam was also laid on to steam-pans for heating the rubber, and to stoves for drying it. When all these means came into operation we soon found them insufficient to meet the still increasing demand; and more power was added and the works extended. I think we were now frequently using from two to three tons of rubber weekly.

The shameful adulteration of the raw material in various ways, and the admixture frequently of an enormous amount of gross matter, such as sand, wood, clay, and other rubbish, gave us a great deal of trouble, and required a considerable amount of steam power, as well as manual labour, to clean it, besides the large pecuniary loss sustained.

I some time since, through the medium of the "Gardener's Chronicle," called attention to the possibility of cultivating the best kinds of caout-chouc-bearing plants in the East and West Indies: the Siphonia elastica of the river Amazon, the Hancornia speciosa of Pernambuco, and the Urceola elastica of Borneo, Pulo-Penang, and other islands. From the best information I have been able to obtain there is every probability of success, and as this substance is now become an article of large and increasing consumption, plantations of these trees may in a few years produce a beneficial return. We cannot look far into futurity, but if by any chance the present source of supply should be cut off or obstructed, another source would be of great importance; at all events the subject should not be lost sight of, and I mention it here in the hope that the suggestion may meet the eye of some one who may be disposed to make the trial. Sir W. Hooker has been so kind as to say that he would at any time render any assistance in his power to parties disposed to make the attempt.

In 1835 I obtained a patent for an improvement in rendering air-beds, cushions, pillows, &c., more elastic and capable of yielding more uniformly to the pressure of the body. When pressure is applied to an ordinary air-cushion, the air contained in it being compressed, forces out the flexible, but comparatively inelastic, material of which the cushion is formed to a state of tension. This strain upon the material is of course in proportion to the degree of pressure employed. I once had the curiosity to put an ordinary pillow between boards into an hydraulic press, to test the weight it would sustain, and was surprised to find that it bore a pressure of seven tons before it burst; so that in fact an air-cushion, although it accommodates itself equally to pressure when in use, it then becomes comparatively hard; yet it still has the advantage of yielding, and accommodates itself to every motion, however trifling. The object of this patent was to communicate such a degree of elasticity to the material which confines the air as to enable it to expand freely, when subjected to the weight of the body; so that the air, being less restrained, could now contribute its elasticity to form a soft and yielding, instead of a hard and tense, surface to the cushion. The air-proof cloth for this purpose was rendered elastic by stretching threads of rubber on a frame to their utmost tension at certain distances from each other, and fixing them at the ends: the cloth was also strained on a board with the rubber-surface upwards; the threads still on the frame were then placed on the cloth and rubbed down; the interior divisions of the cushion were then placed on them; a piece of air-proof cloth of the same size as the former, and furnished in the same manner with rubber-thread, was then laid on, and the whole well rubbed down; the threads were then liberated, and, by applying heat, the cloth was gathered up into a corrugated surface, and the whole cushion rendered elastic and yielding. The specification is given in the Appendix. These articles were highly valued by invalids, but they were apt to get out of order by the frequent gathering up of the thin film of rubber on the cloth, and they were also difficult to repair; and we reluctantly gave up making them. These expanding cushions were a very near approach to Dr. Arnott's "Hydrostatic Bed." As the doctor generously gave this invention to the public, they were soon introduced into hospitals, and kept by parties who let them out on hire. Our trade in sheets for these beds became considerable.

My process of mastication, which had been kept secret for twelve years, was at length divulged by one of my workmen, and, by this time, 1835, had become known to several persons, who, taking advantage of this disclosure, were enabled to make solutions with facility; and hence Mr. Macintosh's patent began to be infringed, so that, in the latter end of this year, proceedings were taken against the infringers, and the action came on to be tried in February, 1836, when a verdict was obtained, establishing the validity of the patent. After this, other infringers, fearing consequences, solicited our forbearance, and no further proceedings were taken, and the patent remained undisturbed during the remaining period of the grant.

In 1837 I obtained a patent for improvements in rendering cloth waterproof. Hitherto, the rubber used for spreading on cloth had been reduced to a state of solution sufficiently attenuated to admit of its being spread without heat in the machine before described. By my present improvement, the quantity of solvent formerly used was greatly reduced, and, in some instances, spread without the use of any solvent at all. The masticated rubber was rolled into rough sheets, and such a proportion of solvent applied to both sides with a brush as would in the end soften the rubber to the desired degree. It was then rolled up together, and allowed to remain covered up till the next day; it was then submitted to the action of rollers of differential speeds until the whole became of the consistence of a smooth and plastic mass.

The rubber, thus reduced to a doughy state, was spread by an apparatus something similar to that before described, except, that instead of a wooden

plank for the bed of the machine, a revolving hollow
iron cylinder was substituted, and kept hot by
steam or hot water; and the coated cloth was
made to pass over a flat iron chamber heated in the
same way, to evaporate the small quantity of
solvent the rubber might still retain.   These improvements are so decisive that they are retained
to the present day, and a great number of these
machines are kept in constant operation.   The
masticated rubber without any solvent has been
spread by these machines; but the spreading
succeeds best when the rubber is in some degree
softened by a small addition of solvent.   (See
Specification in the Appendix.)

In 1838 I obtained a patent for manufacturing
rubber into wide sheets of any required thickness
or length.   The preparation and spreading of the
Rubber was effected by the same machinery as
that described in my patent of 1837, the principal
difference being in the preparation of the cloth
upon which the rubber was spread.   This patent
being for the production of sheets of rubber without cloth attached, it was necessary to prepare the
cloth upon which the rubber was to be spread in
such a manner as to admit of the cloth being easily
detached or stripped off.   This was done by saturating the cloth with a solution of gum, starch, or
glue, and calendering it so as to produce a smooth
surface.   After a sufficient number of coats of the
rubber-dough had been spread upon this cloth to
make up the desired thickness, the whole was im-

mersed in warm water to dissolve the gum contained in the texture of the cloth, when the sheet
of rubber came off with ease. This patent has also
been used to a very great extent, and continues to
be used largely to this day.

In the summer of the year 1838 we were visited
by a great calamity. Our works at Manchester
took fire at midnight, after all the men had left,
and although the building was constructed on what
is called the fire-proof principle, such was the inflammable nature of the large amount of stock it
contained, that all that was combustible was speedily
destroyed, and, what is worse, several of the men
lost their lives by the unexpected falling of some
of the heavy machinery through the arched floors,
supposed to have been secure against such a
casualty. I happened at the time to be in Scotland, on a visit to my late much esteemed and lamented friend and partner Mr. Charles Macintosh.
I hastened to Manchester, and soon witnessed the
scene of devastation, and heard of the appalling
circumstances in the melancholy loss of life of the
poor men who had ventured to stay too long in the
devoted mill.

The walls were standing, and the chimney and
stone stairs, and most of the arched floors, but the
machinery was mostly damaged beyond repair, and
the reservoirs of solutions and solvents, being chiefly
at that time of wood, had of course disappeared.
We never could ascertain the cause of this disaster:
it remains a mystery to this day.

Having plenty of zealous help at hand, the rubbish was soon cleared away ; in the meanwhile, no time was lost in devising means for resuming work. This was the more pressing as our business was yearly on the increase, and the season for the demand of our goods fast approaching. At first it was proposed to bring steam across the street from our partner's cotton-mills by an existing subterranean passage ; but while this was being deliberated upon, the engineer of the works came with the welcome news that the two engines had not sustained material damage, and might soon be got to work. Hands were put on in every department, for the machinery, and builders to restore the mill, and the whole was sufficiently restored to resume operations in an incredibly short time ; and I think, after all, if I remember right, we still did more business than in any previous year, frequently waterproofing from three to four thousand square yards of double textures a day.

In the year 1839 we constructed some very large air-proof vessels or chambers, in the form of flat-bottomed boats. These were intended to supersede the metal pontoons generally used in the construction of military bridges, for conveying troops, ordnance and stores across rivers. As many as twenty pontoons, or even thirty, are often necessary to form a bridge, and as they are twenty feet long—large, slow-moving carriages like timber-wagons being needed to carry them — they greatly

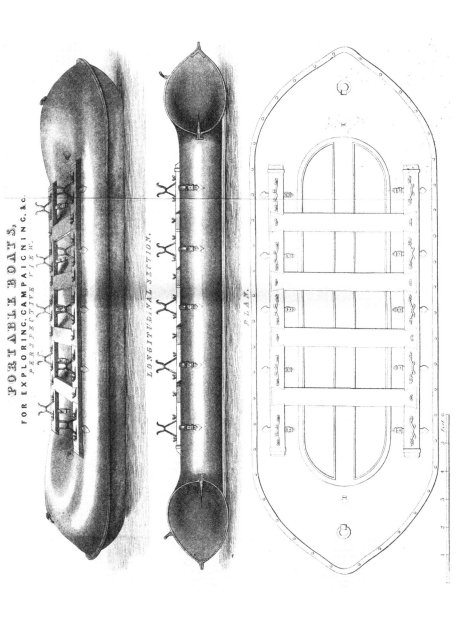

The material originally positioned here is too large for reproduction in this reissue. A PDF can be downloaded from the web address given on page iv of this book, by clicking on 'Resources Available'.

impede the march of an army; so that in the Penin-
sular War, the pontoon train was often in the rear
and not available when required.   The object now
in view, the invention of an eminent military officer,
was to make pontoons more portable, by construct-
ing them of canvas of great strength rendered air-
proof by rubber.

The pontoons when inflated were found fully
capable of sustaining the bridge and the passage of
troops and artillery, and, when exhausted of the
air, could be folded up into a comparatively small
compass, and were light and portable enough to be
carried on one horse.   Each pontoon was furnished
with several openings, having a screw nozzle attached with a proper valve to close it: to these
nozzles a kind of cylindrical bellows was screwed
on; and with one man acting on each pair of
bellows filled the pontoon in from five to six
minutes.

The timber superstructure of the bridge was
formed much in the usual way; duly attending,
however, to the object of its portability on animals,
or small carriages about the length of a field-piece
on its four-wheeled travelling-carriage the pontoons being lashed to the superstructure, maintained their position and proper distance from each
other.   Forty of the foot guards, under the command of an officer, were placed upon the raft sup-
ported by two of these pontoons, and were towed
down the river some distance.   The men were

perfectly at ease on it.   The Duke of Wellington, in
order to judge of its stability, and the room afforded, caused the party to sit down, and even to lie
on it as well as to stand upright.   The raft was
then towed to shore and the men were landed.   I
afterwards attended some experiments at Woolwich
with these pontoons in the presence of Lord
Vivian, who appeared to be satisfied with their
performance, and made minute inquiries as to
their endurance, and the means of providing
against perforations by shot or accident.   He also
had a piece of ordnance embarked on it, and a
posse of men to handle it.   All went off quite
satisfactorily.   These pontoons were employed in
pontoon exercise for some months, and the inventor of them was not slow in acknowledging the
assistance we had rendered him in carrying out his
plans.

In 1840 we found it necessary again to increase
our means of production by the addition of 'more
power, and of new buildings and machinery.   The
machinery was now on a still much larger scale:
the rollers for breaking down the raw material
and cleansing it were modified in form, and the
heavy rollers for crushing the rubber for the
masticators were of a strength suitable to the preparation of the charges of rubber — for the mon-
ster masticator was for the first time put to work
(a perspective sketch of this machine is annexed) —
each charge required for it amounting to from 180
to 200 pounds.   For the formation of this large

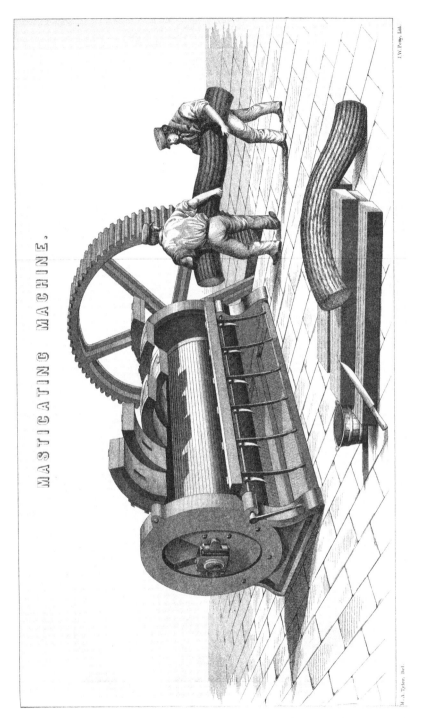

MASTICATING MACHINE.

The material originally positioned here is too large for reproduction in this reissue. A PDF can be downloaded from the web address given on page iv of this book, by clicking on 'Resources Available'.

mass into square blocks, new moulds had to be
made, with means to apply suitable pressure.
These were now six feet long, twelve inches wide,
and six or seven inches thick. As before stated,
these required large cutting machines which enabled
us to cut sheets of any thickness required by our
customers; and such was the perfection to which
these processes were brought, that for particular
purposes sheets were cut as thin as the eightieth
part of an inch, and of such a size as to be used
without joining for purposes where hitherto it had
been requisite to join them.

A patent had been taken out for raising sunken
ships, and we were now called upon to make some
very large air-tight cylinders capable of sustaining
great pressure : these were much larger than the
pontoons, and in those days tried the skill of our
workmen to make them sound, and of the forms
required. This, however, was accomplished, and
with these, experiments were made to raise some
sunken vessels. To effect this, the cylinders were
collapsed and sunk to the lowest part of the vessel
to be raised, and attached to each other by chains
underneath its bottom. The cylinders were then
filled with air by means of hose-pipes, and air-
pumps worked by steam.

The theory of the inventor appeared to be right
in principle and very promising; the cylinders
bore the pressure, and there seemed to be nothing
to prevent success. Such, however, was not the
case: the great difficulty proved to be the getting

chains under the vessel, and placing them in the required positions. The construction of these large receptacles, and the various schemes we were called upon from time to time to assist in carrying out by means of our manufactures, tended constantly to improve our manipulations, and add to our experience in adapting it to a vast number of applications both of such as suggested themselves to our own minds as well as those that were adopted by others.

As it is my intention to confine myself as much as possible to a personal narrative, I shall only mention here, that although several parties had now entered into the rubber manufacture, and some of our hands left us, and carried with them what they had learnt whilst in our employ; and although some of these competitors boasted of large means, and made a formidable array of trustees, directors, botanical adviser, manager, auditors, solicitors, bankers, secretary, &c., they did not much injure the progress of our business. The proceedings, therefore, of these parties were not such as to cause us any great amount of anxiety; nor did we suffer any material diminution of our sales thereby, although we suffered in some degree by the depreciation of prices in some articles.

These parties, as well as ourselves, were destined to witness a falling off in the waterproof department, from other and very different causes, for although our trade in these articles continued good for another year or two, yet we felt a gradual decline in the demand, and a universal depression in the sale of them throughout the country.

Several concurrent circumstances contributed to bring about this state of things. Those travelling on the outside of stage coaches were the most numerous class of our customers for waterproof articles, and those stage coaches were fast disappearing before the superior speed and accommodations afforded by the railway system ; here roofs and protection were afforded for travellers formerly exposed on the top of a coach to all the changes of our variable climate, greatly needing then the protection we afforded them ; these now needed it no longer, or to a comparatively limited extent. In the neighbourhood of large towns, the introduction of omnibuses in a great measure superseded the short-stage coaches. Gigs and light open carriages were also much less used. All these changes occurring almost simultaneously, necessarily operated to lessen the demand for waterproof clothing. But last, although perhaps not least, the doctors spread a universal outcry that these garments were so unhealthy, that no one ought to expose himself to the hazard of wearing them ; forgetting that their predecessors, and some of the elders among themselves, were constantly attributing many of the numerous " ills that flesh is heir to," to a drenching cold rain and a wet skin, and that a garment to provide against such unhappy consequences was a desideratum which had long been sought for in vain, and which none, perhaps, were more earnest than themselves to obtain.

It would have been more consistent with the professional knowledge of medical men, if instead of decrying these garments, which they admitted were an effectual protection from rain, they had given to their friends any precaution they thought necessary, and particularly against their fitting so close as to confine insensible perspiration, which the wearer himself soon discovers to be unpleasant. All this is pretty well understood in the present day, and loose-fitting waterproof garments being made much lighter than formerly, they are now no impediment to the pedestrian, and are consequently in more general use than at any former period, and I believe no longer objected to by the faculty as injnrious to health.

Several patents were taken out by the late Mr. W. Brockedon, for various modes of manufacturing a substitute for corks and bungs, of which rubber formed a component part. With the early part of his experiments, we had nothing to do, but ultimately he applied to us for assistance in carrying out in succession the several improvements he had projected.

The core of these corks and bungs was originally formed of woollen felt, and coated first with solution and then with thin sheet rubber: these were found too expensive for most purposes, and recourse was had to a structure of cotton for a core. The manufacture of this core underwent several

NAUTICAL ARTICLES.

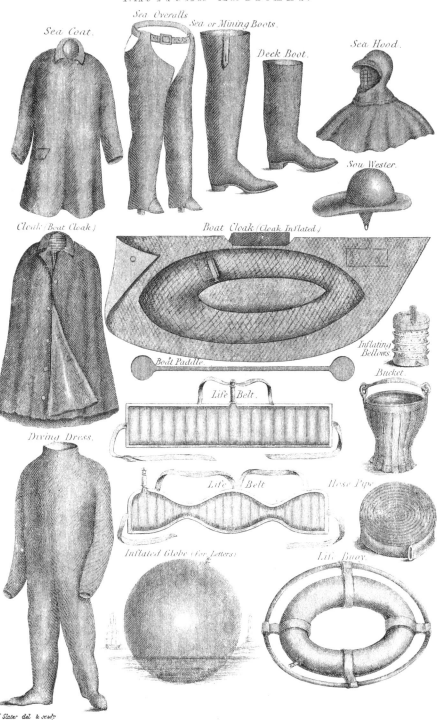

Sea Coat.

Sea Overalls.

Sea or Mining Boots.

Deck Boot.

Sea Hood.

Sou Wester.

Cloak (Boat Cloak)

Boat Cloak (Cloak Inflated)

Inflating Bellows.

Boat Paddle.

Bucket.

Life Belt.

Diving Dress.

Life Belt

Hose Pipe

Inflated Globe (For Letters)

Life Buoy.

Slater del & sculp

successive improvements, till at length, by the
invention of some very beautiful machinery, Mr.
Brockedon brought this core into great lengths in
a cylindrical form, from which corks could be cut
of any required length.   The cotton constituting
this core was so placed and bound together as to
yield sufficient elasticity without being too easily
compressed.   The corks so formed were also coated
with solution, and covered with thin sheet rubber,
forming an article every way apparently suitable
to the intended use.   Much time, great ingenuity,
and a large outlay were absorbed in this undertaking, and for a time everything promised well.
Of course wine merchants were expected to be
the largest consumers of these corks, and accordingly they were introduced to various houses
engaged in that businesss.   Many were pleased to
have a firm clean stopper, instead of the faulty, perforated, dusty things they had hitherto used ; but
they had to undergo severe testings.

As nearly as I can rememember, about 1840, a
gentleman from Paris gave me a small kind of
doily, or small piece of cambric, covered on one side
with rubber of a dingy yellow colour, on which
was the print, I think, of a female.   He also gave
me a map printed on rubber without cloth, which
would expand by the elasticity of the material.
These were the only articles, as far as I recollect,
that I had yet seen that were not of English or
French manufacture.   The yellow tinge in the

rubber on the cambric was produced by the use of sulphur, and the appearance of the article by no means good.

Mr. Brockedon continued indefatigable in the endeavour to improve his stoppers : the colour of them, which was that of sheet-rubber, was objected to, and he was anxious to get a better surface, and had recourse to many expedients with that view, in which from time to time I assisted him, and at length their form and the surface and general appearance were greatly improved. In the meantime they had to be subjected, for some considerable period, to various trials, as to their efficiency amongst the wine merchants, for soda-water, beer-bottling, bungs for fruit preservers, gunpowder canisters, and various other similar uses.

Although the waterproof trade had somewhat declined, yet we still did a considerable business in that line, and the air-proof department, I think, never flagged much. Life-preservers became more known, and their utility had been proved in the saving of life. The demand for diving-dresses and the tubing for them increased as well as the trade in hose-pipe, fishing-boots, &c., and the trade continued good for fine cut sheet-rubber, and blocks for stationers, &c. The trade in solutions had now become considerable ; not only for rendering the seams of garments waterproof, and for shoemakers, but as a cement also, which for some purposes it is particularly well adapted, possessing this peculiarity — that it remains adhesive when dry, and two surfaces coated with it, when brought into contact, immediately unite, requiring only a little pressure or rubbing with the hand. This cement will be used extensively when more known.

Some time in the early part of the autumn of 1842, Mr. Brockedon showed me some small bits of rubber that he told me had been brought by a person from America, who represented himself as the agent of the inventor: it was said that they would not stiffen by cold, and were not much affected by solvents, heat, or oils. Mr. Brockedon told me that the mode of manufacturing this rubber was a secret, and that the agent who had shown them to him, declared himself totally ignorant of it ; all he knew was, that it had been done by some new solvent which was very cheap in America, and that he wished to find parties who would purchase the secret in this country, and asked Mr. Brockedon to give him any introduction he could to the trade. Mr. Brockedon gave him the names of Macintosh and Co., and the representatives of the Caoutchouc Company, (and, I think, at least one other) with whom the party also left specimens. Macintosh and Co. told the agent, that as he could give no information, they could not judge of the merits of the invention, as it might be easy to make these small specimens, whilst difficulties might be found in its application upon a large scale ; nor could they judge whether their present appliances, which had been very costly, would be suitable for this manufacture, or whether any modification of their present plant would answer, or if the modes were so altogether new as to require a fresh outlay. Under these circumstances, as they could not act in the dark, they recommended the taking out of a patent, when a clear comprehension of the whole could be obtained, and the invention openly dealt with according to its merits. This course was approved of, and the agent said he would immediately advise his principal to take out a patent.

In the meantime, Mr. Brockedon had become sanguine about his stoppers, as nothing at present appeared to make their success doubtful, providing they did not in time impart any flavour to the wine. As he had now seen that rubber could be so prepared to be unaffected by cold, he became desirous of imparting that quality to his stoppers, and immediately commenced experiments to see whether he could find some way of producing the same condition in the rubber for the purpose. I believe he pursued those experiments for a considerable period, and at one time told me he had succeeded; but in this he was mistaken. His preparations would still stiffen by cold.

Mr. Brockedon cut, from two of the small pieces which had been given him, two very diminutive bits, and gave them to me; one bit was about one and a half or two inches long, and one inch and a quarter wide, and perhaps about the twentieth of an inch thick : this bit, on the exterior, was of a

dirty yellowish grey colour, and a little dusty
powder upon it; when cut across, the cut edge
appeared of a dark colour : the other bit was of a
dark reddish brown colour throughout, with a clean
surface without any dusty appearance. The first
named had a slight smell of sulphur, and I thought
that a little sulphurous powder had been rubbed
on to mislead. On stretching them out thin, I
observed they were both quite opaque. I found
they were, as stated, not affected by cold.

About this time — that is, the latter end of the
year 1842 and the beginning of 1843, — I was
engaged in some experiments for divesting rubber
of its adhesiveness, particularly the surface of solutions of rubber spread upon cloth to render it
waterproof, with the view of avoiding the necessity of employing a double texture for this purpose,
according to Mr. Macintosh's principle. Cloth
so prepared has since been called " single textures,"
and used very extensively for garments and
other coverings. Being engaged in these experiments, I determined at the same time to try
whether I could not also find some means of
preventing rubber from stiffening by exposure to
cold. The idea was not new to me; I had attempted it nearly twenty years before, and thought
I had succeeded ; but the goods to which my preparation was applied would not bear the test of
time; they were spoiled ; and the whole resulted in
a rather severe loss.

Finding now that this object appeared to have

been somehow or other effected, and therefore demonstrated to be practicable, and as it was my particular department to keep up the quality of our manufactures, and to maintain our standing and the position our goods had attained, I set to work in earnest, resolved, if possible, not to be outdone by any.   As I advanced, I spent all my spare time to attain the two objects I had in view.

I succeeded in producing beautiful drab and tan-coloured surfaces, perfectly inadhesive, by combining silicate of magnesia, and compounds of this silicate, with fuller's earth, whiting, ochre, and other earths, in all of which the silicate of magnesia formed the foundation and predominated. For a dark colour, I employed asphalte, both the natural mineral kinds, as well as that obtained by boiling down coal-tar to the state of asphalte. With this asphalte, I also sometimes combined plumbago, and by these means produced an unadhesive black surface.   Since that time, a taste for a more glossy article prevailing, varnishes have been added.   In carrying out these improvements, no new machinery was needed, the rubber and compounds being made, and the straining and spreading effected, by the means before mentioned.

During my progress with these experiments, I gave attention also to my other object; and having nearly brought the former to a close, I could give to the latter more special attention.   The little bits given me by Mr. Brockedon certainly showed me for the first time that the desirable change in the condition of rubber of not stiffening by cold had been attained, but they afforded no clue to the mode by which it had been brought about.

As the knowledge of this improvement in the qualities of rubber soon spread abroad, several persons began to make experiments, in order to discover means to effect the same ; but as there were very few who at that time had any considerable experience in the treatment of rubber beyond the ordinary applications for water-proofing purposes, and the construction of pneumatic articles, they were not very likely to succeed in an attempt which required, besides an experimental knowledge of the subject, unwearied application, and the closest observation of the effects of every new addition to the compounds, and of the proportions indicative of improved results — nor were high temperatures employed by such persons.   Mr. A. Parkes, for instance, was not much acquainted with the rubber manufacture, and therefore, after he had seen my specification, and the change that had been effected by sulphur and heat, went immediately to work as a chemist, and tried the effect of the whole round of sulphurous combinations with the rubber, and for a time gave up the pursuit, but being urged to continue his experiments, he at length succeeded, by purely chemical means, in effecting a change in rubber very closely resembling vulcanization.

It is a singular fact, that although sulphur had long since been compounded with rubber by myself in this country, and in America by others, yet that its true value for producing such a result had never been even dreamt of by any of us. I made no analysis of these little bits, nor did I procure, either directly or indirectly, any analysis of them. In making my experiments, I depended entirely and solely on my own exertions, having some confidence in an experience of upwards of twenty years of unceasing application to the manipulation with my own hands of the substance I was dealing with; and it having been shown that rubber could be so made as not to stiffen by cold, I devoted myself to the discovery, if possible, of some mode by which this property could be imparted; and I considered the small specimens given to me simply as a proof that it was practicable. It had been my constant practice to make my experiments alone, in my private laboratory at Stoke Newington, into which on these occasions no person entered but myself. I consequently lighted my own fires, and did all the labour myself; and having constantly to attend to the general business in London, and to answer innumerable inquiries, both verbally and by letter, besides attending consultations with parties taking out patents, or others requiring information in the application of our materials to their different purposes, I was obliged in a great measure to pursue my experiments after the hours of business, and was frequently employed

till midnight in these labours. I think it necessary to mention this, because it has been alleged
that nothing could have been less difficult than
such a discovery, and that any one acquainted
with the substance might have accomplished it
with great ease. After a discovery is once made,
it is generally said to be easy of attainment; any-
body might have done it, the thing was so simple.
According to Milton, an idea of this kind occurred
to some as far back as the council in Pandæ-
monium: —

> " The invention all admired; and each how he
>   To be th' inventor missed, so easy it seem'd,
>   Once found, which yet unfound most would have thought
>   Impossible."

But to resume, I knew nothing more of the composition of the small specimens given me by Mr.
Brockedon than what I or any other person might
know by sight and smell. In such a pursuit I was
obliged to grope a good deal in the dark; for although my former experience, such as heat, masticating, rolling, and the use of solvents, served
as landmarks, still my experiments could only be
made pretty much at random, not at all knowing
what the result of any of them might be. I remember well that I had from the first a strong
impression that the rubber underwent the change
either when in a state of solution, or when greatly
softened by heat, and these two points I kept
pretty much in view. I compounded with the

rubber in one or other of these conditions an
almost endless round of matters of all kinds.

I treated these with the rubber separately, and
in innumerable combinations. In some of them I
included sulphur, and employed heat in almost
every case, without regarding what degree of heat
I used. I have, in an early part of my narrative,
stated how useful I found it to employ high temperatures, and attribute my final success in my present
pursuit to the habit I had acquired in its use,
particularly when I thought it necessary to reduce
the rubber to the softest or most plastic state without solvents. When making the compounds with
solvents, I first dissolved the rubber, either making
the solution of a thin consistence, or of the consistence of dough, and then mixed or worked up the
other matters with it. When I wished to expedite
their drying, I sometimes laid them on a small
metal plate, heated over a chamber lamp, or larger
plates heated by the fire, and at other times I
submitted them to the heat of an oven. As before
mentioned, I frequently employed sulphur in the
compounds, but they were not at all improved by
it, not having yet by any chance used heat sufficiently high when drying them to produce the
change ; or if the heat was sufficient, which it most
likely sometimes was, the compound became dry in
too short a period to affect it (as I removed the
pieces as soon as they were dry) in any way to
attract my observation, having nothing to guide
me in the object in view. I therefore for a time

relinquished the use of sulphur in most of the compounds as useless, and pushed on with other matters, still feeling a conviction, as I think most would naturally have done, that the rubber must necessarily undergo the change in its constitution, whilst either very soft and plastic, or in a state of solution; however, after trying an endless round of mixtures in this way, with and without heat, I still failed of success.

Whilst looking over some of my former experimental scraps, I saw in some of those containing sulphur, variations I could not at the time account for: some portions were different to others in the same specimen, which for the present I could not comprehend, although I well knew afterwards. I resolved now to take another course; I dissolved sulphur in oil of turpentine, and finding the solution proceed slowly, I raised the temperature of the turpentine to the boiling point (316°), and then it took up the sulphur freely. Rubber dissolved in this solution was good in some respects, but produced a weak material, and did not come up to my wishes.

I spent all my spare time for months with these experiments; my habit was to dispatch them quickly, making them very small in bulk, throwing aside some thousands of trial scraps, and selecting and keeping for inspection any that appeared promising. During the winter months I generally found the weather cold enough to test my scraps;

but as the spring and summer came on, I employed
ice, purchased from an ice-cart which passed my
gate every morning (from Southgate on its way to
town), and many an earnest and careful examination have I made in a morning of the scraps in
the ice.

My experiments now had become very interesting; I had certainly produced in some of my scraps,
or portions of some of them, that condition of
rubber which I afterwards called the " change."

I cannot of course now say in which scrap or at
what particular moment I struck out the first spark
of the " change," because I had a great number of
scraps under my hand at one time, and I also
found that when exposed even to ice, the stiffening
effect of cold was not always immediately apparent;
and I was thus during my comparative ignorance
of the nature of my discovery frequently misled
and disappointed from two causes: first, what it
was in the successful compounds that produced the
effect; and secondly, I was ignorant of the importance of the degrees of temperature employed, as
well of the like importance of the period to which
any specific compound required to be submitted to
heat.

These points could be ascertained with exact-
ness only by time and vigilant watching, and this
I began immediately to set about; and as the law
allows a patentee six months to work out his discoveries before he is called upon to enrol his specification, I applied for and obtained a patent for

my inventions, which passed the great seal on the 21st November 1843.

I shortly afterwards received an intimation that my late lamented friend and partner Mr. Macintosh, who had been some time in very indifferent health, was now seriously ill, and had expressed a particular wish to see me. I started immediately for Glasgow, but on arriving at Lancaster, I received an intimation that I was too late, and that his death was momentarily expected, which happened on the 25th of July, 1843. I have never met with any person for whom I entertained a greater esteem, nor could any two persons in business act more cordially nor with more frankness, and I believe mutual esteem, than subsisted between us.

Mr. Macintosh was a Fellow of the Royal Society, and member of several scientific societies; as a chemist he enjoyed an European reputation, and had long been largely engaged in chemical operations in Scotland. Some of these, such as the alum works at Campsie in Stirlingshire, were carried on upon a landed property which he possessed.

An interesting biographical memoir of him, written by his son and successor the late Mr. George Macintosh, was printed for private circulation, but was never published.

I found that, when submitting the compounds containing sulphur to heat, it was necessary (as before observed), after ascertaining the temperature

that suited any compound, to find also the period
of exposure to the heat that produced the best
result.   Until I noticed the necessity for this par-
ticularity I was often sadly perplexed; as the same
compounds exposed to the same temperature were
sometimes good and sometimes bad in practice, the
variation is from one to six or seven hours, or
more.   All the way through these experiments for
producing the " change," I had no other guide
or course than to watch for any promising appearance in any of the scraps, and to improve upon
them ; but I now know I was frequently thwarted
by my want of information as to what cause
different appearances were due; and particularly
in regard to the temperature I employed, which
was somewhat at random, knowing how freely
I could use it within certain limits without injury.

A thought now occurred to me that in the end
proved extremely valuable.   Revolving in my mind
some of the effects produced by the high degree
of heat I had employed in making solutions of
sulphur and rubber, as before stated, in oil of
turpentine, it occurred to me that as the melting
point of sulphur was only about 240°, which I
knew would not be injurious to the rubber, it
would be well to see what would ensue on immersing a slip of sheet rubber in sulphur at
the lowest melting point.   I accordingly melted
some sulphur in an iron vessel and immersed
in it some slips of cut sheet rubber, about half
an inch wide, and about one sixteenth of an

inch thick.   After they had remained some time
I examined them, and found the surface had assumed a yellowish tan colour.   I immersed them
again ; and on withdrawing them the second time, I
cut one of them across with a wet knife, and found
that the rubber was tinged of this tan colour to a
considerable depth.   I immersed them again; and
on the third examination I found the tan colour
had quite penetrated through the slip.   This was
strong evidence that the rubber had freely absorbed the sulphur, and I fully expected to find
that these slips were now " changed," but in this I
was greatly dissappointed, for on applying the tests,
I found that not the least " changing " effect had
been produced.   I now replaced them and raised
the temperature of the sulphur, and allowed them
to remain a considerable time ; and on withdrawing
one of them the fourth time, I found, to my great
satisfaction, that it was perfectly " changed," retaining the same tan colour throughout ; the other
slips remained in the sulphur whilst this examination was going on, and on withdrawing them, I
found the lower end nearest the fire turning black,
and becoming hard and horny, thus at once and
indubitably opening to me the true source and
process of producing the "change" in all its states
and conditions, and in all its pure and pristine sim-
plicity.

I need not say that I was greatly surprised and
pleased with these results, and the excitement I
then felt is somewhat revived whilst I am writing

the account of this stirring incident in my operations.

The value of this experiment with melted sulphur was incalculable, as it at once discovered to me the true principles on which the "change" was based. Although I had produced in my earlier experiments precisely the same result on as much rubber as any of the compounds happened to contain, yet there had always been comprehended in them other matters, and particularly solvents had been used; it was now demonstrated that sulphur and sulphur alone, blended with the rubber, and acted on by heat at a proper temperature and for a relative period of exposure to its influence, was the sole cause of this extraordinary change in the substance.

I believe I may confidently affirm that it was not previously known even that rubber would absorb sulphur, and no degree of chemical knowledge, no analogy of reasoning on the relative nature of the two substances would ever have suggested or anticipated the possibility of such a result as the "change" from such combination. If I had known the simple mode by which this result could be produced, I might have made the discovery at once, without spending months of toil and thought upon it.

I now saw my course straight before me; my experiments had shown that the rubber and sulphur must first be blended, and all my former appliances, I could see at a glance, would serve me

in my new career. Sulphur could be blended by
rollers and masticators ; the blended material could
be reduced to a state of solution by any of the
usual solvents, or to a state of dough, and spread
on cloth by the machinery of my previous patents
for waterproofing, or on sized cambric for sheets.
Blocks of the masticated compound could be pressed
in moulds and cut into sheets, and into every
variety of form and size that the new uses of the
" changed " rubber might require ; and last, though
not least, ordinary cut sheets of pure rubber in the
form they came from the knife, or made up into
other forms, could be immersed in a sulphur bath,
and changed to any required degree, from the
softest and most elastic up to a state of hardness
similar to horn, and capable of being wrought with
carpenters' tools, or turned like ivory and ebony in
a lathe.

This mode of producing the "change," so unique
in many respects, is particularly so in this, that
articles manufactured of pure rubber with which
no sulphur has been blended may be changed in
the sulphur bath ; take, for instance, a native South
American shoe, immerse it hot in the bath, at the.
lowest temperature at which sulphur will melt, say
240°, and let it remain until it has absorbed enough
sulphur ; then gradually raise the temperature to
265° or 270°, and in the course of about an hour
or something more, according to thickness, the shoe
will be perfectly " changed."   I know of no other
mode by which this could be done.   This sulphur

bath process is also one of the best I know of for
hard india rubber.   The particulars for obtaining
it in different states or degrees of hardness are
given in my specification in the Appendix.

The enrolment of this document became due on
the 21st of May, and for some time previous to
that day I made every exertion, and took the greatest pains to render it in all respects as complete as
possible.

I called on Professor Graham, and the late highly
respected Mr. Arthur Aikin, in April, 1844, and
left with them specimens of the results of my
experiments, and requested them to repeat them,
and to make themselves masters of the different
modes of operation, in order to their assisting
me in drawing my specification, which they did,
and entered fully into the points submitted to
them.

When this document was drawn, I also obtained
the assistance of Mr. Carpmael, who carefully pe-
rused it, and fully approved of the manner in which
it was drawn ; it was finally settled by the late
Mr. Cowling, and then duly enrolled.   I think I
might venture to state, not boastfully, but as a
matter of fact, that there is not to this day, 1856,
any document extant (including those referred to in
it) which contains so much information upon the
manufacture and " vulcanization " of rubber, as is
contained in this specification.   If any of my
readers can point out such a document, I shall feel
much obliged if they will inform me of it.

Nearly all my experiments had hitherto been done on a small scale at my private laboratory, but my operations were now transferred chiefly to Manchester; where, with slight modifications, the machinery already in our works was sufficient to enable us to start immediately with the processes contained in the first portion of my patent for obviating the adhesiveness of manufactured rubber, not only to the surface of single textures, but to various other purposes, and, among the rest, to Mr. Brockedon's stoppers, which were no longer adhesive nor of an unsightly dark colour, but were now made of the colour of cork, and with a smooth even surface, adding greatly to the neatness of their appearance, and removing other objections that had been made to them. Although, for the reasons before mentioned, they did not come into general use, yet, in this new form, they continue to be manufactured.

In preparing to carry out the new process on a large scale, it appeared desirable to give the material a more definite name than "the change," which I had adopted for the purposes of my specification; and, whilst discussing the subject amongst my friends, Mr. Brockedon proposed the term " VULCANIZATION," and, as no better suggested itself, I determined to adopt it, and it is now known by that name both in Europe and America.

It owes its derivation to the Vulcan of mythology, as in some degree representing the employment of sulphur and heat, with which that mythological personage was supposed to be familiar. I shall, therefore, of course use the term in the remainder of my narrative.

I had now to consider the different modes described in my specification for blending the sulphur with the rubber. The modes had to be adapted to the different purposes to which the material was to be applied, and then the proportions of the compounds for the various objects had to be ascertained; and next, the most eligible means of obtaining the required temperature for vulcanizing each of them, and this also in reference to the uses for which they were designed; as well as to keep in view the action sulphur might have upon fabrics composed both of animal and vegetable fibre, when the compound of rubber and sulphur spread upon them had to undergo the high temperature required.

I found, as before stated, that when rubber is blended by means of the sulphur bath, the blending and vulcanizing may be generally carried on in a manner simultaneously. The absorption of the sulphur takes place at about 240°, and then, by raising the temperature to 270° or 280°, and allowing the rubber to remain in it for about an hour, it becomes " vulcanized." This is certainly the most simple mode, and is also as effectual as any.

When I used the sulphur bath for blending, I removed the rubber as soon as it had absorbed enough sulphur, and before any vulcanizing effect had been produced, I then dissolved it or made it into dough and spread it on cloth, and vulcanized it.

Another convenient mode of blending thin sheets or films I found to be by heating the surface of the rubber, and then rubbing on some finely powdered sulphur, which I found to be amply sufficient for the purpose when vulcanized in steam, as hereafter described. The most convenient mode, however, upon a large scale, is to blend by means of rolling and masticating the two substances together, and then pressing it into moulds to be cut up into sheets and other forms, or to be rolled out and reduced by solvents to a state of solution or dough, and then to carry on the operation by the spreading machine or rollers. The other modes mentioned in the specification answer very well, but these are the best in practice. I may mention here, that although there are comparatively very few colours that will bear exposure to heat in contact with sulphur, yet we soon produced good colours in red, green, black, and white; and here, the process stated in my patent for depriving the material of any excess of sulphur came to my aid, as, when rubber is vulcanized, there arises an efflorescence of sulphur on the surface, which is detrimental to the clearness of colours; the desulphuring process entirely prevents this, and is extensively employed in our manufactures. The white has been used for printing on, and beautiful impressions have been taken from wood and steel engravings, and also from ordinary type or letter-press printing.

My first apparatus for vulcanizing at our works consisted of an iron pan large enough to take a sheet of cut rubber ; this pan was set in brick-work, and heated by a fire underneath ; we found it im-possible to keep the sulphur of an equal temperature in all parts of this vessel ; we were, however, able with some dexterity to vulcanise rubber in it for some of the first uses to which the new material was applied ; one of these was for the formation of an elastic packing between the longitudinal sleepers and the iron rails on the Great Western Railway ; a quantity of vulcanized sheet rubber, about three sixteenths of an inch thick and seven inches wide, was supplied for this purpose, and I believe the effect produced was considered a great improve-ment ; but the expense was thought too great to justify its adoption.

I have a piece of rubber vulcanized in this pan on this occasion still in my possession, and although twelve years old, it is to this day a perfect specimen of the pure vulcanization of rubber.

Hitherto, when I had vulcanized in a high-pressure steam boiler, it was in a very small apparatus, such as is found in laboratories. I now had one con-structed of a more considerable size, and of extra strength, that I might not fear explosions at very high temperatures, which I thought it would be necessary in some cases to use.

The operations carried on by means of this

boiler were important, and were destined to establish the superiority of this mode of submitting the compounds of rubber and sulphur to heat over every other mode for many of the most important applications of the new material, especially for articles of any considerable bulk.

Thin sheets spread on cloth, or fabrics coated with thin films, can be vulcanized in dry atmospheric heat by an oven or stove. Although stoves may be used for some purposes advantageously, yet there is great difficulty in producing in a stove or chamber of considerable dimensions that equal and uniform temperature in all parts of it which vulcanization absolutely requires; hence the practice in these cases of keeping the goods in a constant rotatory motion; and then with great care and vigilance stoves can be made to answer the purpose.

It will readily be seen that the same objections do not apply to the action of heat when steam is employed under pressure. One great advantage arising from this mode is, that little or no escape of sulphur can take place, as is the case in dry heat, so that the proportion of sulphur may be limited to any extent, even as low as one and a half or two per cent, which is sometimes a great advantage, as little or no efflorescence ensues. I believe it to be utterly impossible to do this by any other mode, certainly not by dry heat.

Another advantage of equal importance is, that the heat is always uniform in every part, and capable of being regulated to the greatest nicety and held there; large masses can be thus vulcanized; solid twelve-inch cubes have been done at our works, vulcanized throughout. Sheets fifty yards long and fifty-six inches wide were now made, and also sheets of large dimensions, ten feet square, and nearly three eighths in thickness, for the use of Government; these are called ship-sheets; the application of these sheets was suggested by the late ingenious Mr. John Cow, master boat-builder in one of her Majesty's dockyards (he also constructed portable boats for the Arctic voyagers, and others for the landing of cavalry; they consisted of frame-work, covered with strong canvass, rendered securely waterproof with rubber). Before the application of these ship-sheets, it was necessary to bring a large steam vessel into a dry dock before anything could be done to her bottom, in case of leakage, or to repair a pipe or valve, or to examine, from the interior of the vessel, anything that had happened to her, however apparently trifling, causing a heavy expense, and much loss of time. These sheets were let down by ropes over the sides of the vessel, and brought over the part requiring examination or repair; the pressure of the water against the elastic sheet stopped the leak, and the ship could be pumped dry, and a pipe renewed, and shot holes or leaks stopped with economy and dispatch.

I had, as before alluded to, foreseen that it might be desirable in some cases to deprive the

NAUTICAL ARTICLES.

Perspective View of Pontoon.

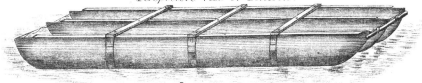

Plan of Pontoon.

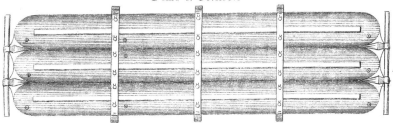

Section of Pontoon.

End Elevation of Pontoon.

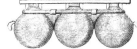

Life Boat Tube.

Raft Tube.

Watching Buoy.

Ship Sheet to stop ingress of Water in case of accident at Sea.

Pontoon Bellows.

I. Slater del & sculp

rubber after "vulcanization" of any excess of sulphur it might retain, and I fully succeeded in effecting this very desirable object, by submitting it to a strong solution of the sulphate of soda or potash at a temperature of about 200°. This process enables me to extract the sulphur to such an extent as not to leave the least appearance of it; restoring the rubber to its natural colour and semitransparency, yet still retaining its vulcanized properties. This has given rise to much speculation; some chemists consider that there is no real chemical combination of the rubber and sulphur at all, but that the change takes place simply by contact of the two substances, under certain circumstances and conditions (some cases very similar are known), or that some new molecular arrangement takes place; but hitherto no definite conclusion has been arrived at which will satisfactorily account for so extraordinary a transformation. The specification of this patent soon appeared in the scientific periodicals, and the process of pure "vulcanization," and the use of the sulphur bath and vulcanizing by steam, was speedily adopted in France, and the latter almost immediately and very extensively in the United States of America.

I come now to the melancholy task of recording the death of another of my partners, Mr. Hugh Hornby Birley, who died on the 31st July, 1845. He took great interest and a very active part in forwarding the views of Mr. Macintosh, both at the commencement of this manufacture, and during his after life in promoting and personally superintending it, and many valuable suggestions and improvements owe their origin to his devotion to our concern; he also bestowed great attention to the construction of both buildings and machinery, when extensions became necessary. He was a most estimable man in all the relations of life, and with uprightness of character, he possessed a calm and dignified amenity, which endears his memory doubtless to many, and especially to me.

It will be easily imagined that the introduction of so new and useful a material as vulcanized rubber would soon attract the attention of ingenious persons. Without taking the trouble of ascertaining what has been done in patents of applications since the Great Exhibition of 1851, I may mention that there had been then upwards of fifty patents taken out by different persons adapting and applying it to their various purposes. The first patentees who applied for vulcanized rubber were Messrs. Perry & Co., in December, 1844, for the formation of their patent Elastic Bands, which have been so extensively brought into use; and in the same month thick sheeting was supplied to Messrs. Betts & Co. Washers and packing for steam pipe joints and other uses were at the same time extensively distributed to various engineers for trial, and in February following orders came in daily for these purposes, and went on increasing, and have never ceased. In May, 1845, engine-valves were sent out and success-

SURGICAL & HOSPITAL ARTICLES.

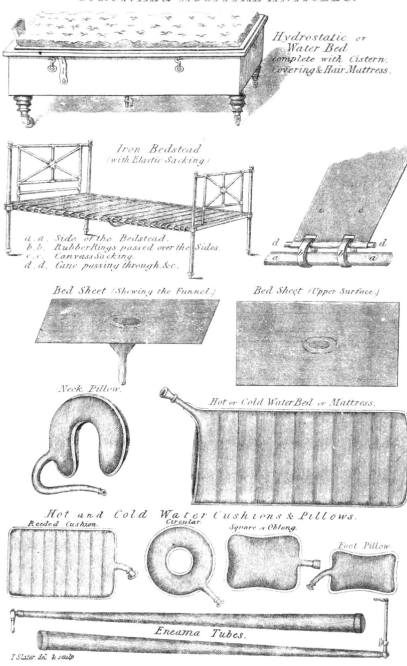

Hydrostatic or Water Bed complete with Cistern, Covering & Hair Mattress.

Iron Bedstead (with Elastic Sacking.)

a.a. Side of the Bedstead.
b.b. Rubber Rings passed over the Sides.
c.c. Canvass Sacking.
d.d. Cane passing through &c.

Bed Sheet (Shewing the Funnel.)

Bed Sheet (Upper Surface.)

Neck Pillow.

Hot or Cold Water Bed or Mattress.

Hot and Cold Water Cushions & Pillows.

Reeded Cushion.   Circular.   Square or Oblong.

Foot Pillow.

Eneama Tubes.

I Slater del & sculp

fully applied.   In June, we sent nearly 4000 lbs. of corrugated vulcanized sheets to the South Devon Railway Company.   In the same month we supplied vulcanized printers' blankets ; billiard cushions followed, spongy vulcanized rubber for musical instruments ; then hose-pipe, tubing, and surgical bottles, besides experimental matters to various persons ; all these things were done during the *first years* of my patent, and I believe nearly all were of pure vulcanized rubber, although for part of it some of the adulterating mixtures mentioned in my specification were introduced to lessen elasticity, &c.   During the following year, 1846, we made vulcanized blankets with spongy surfaces, — gig and carriage springs, solid rubber wheel-tires *, draw and buffer springs for railway carriages, artificial leather, with japanned surfaces, pump-buckets, gas-holders, artificial lea- ther for carding-engines, buoys and air-vessels, inflated wheel-tires, vulcanized thread.   Then followed trouser-straps, vest back straps, swimming- belts, socket pipe joints, printers' rollers, boots, over-shoes, machinery bands, spherical valves, horse boots, and knee caps, furnishers for calico printers, &c. &c.   We also made a very useful material for the soles of boots and shoes by combining asphalte

---

* These tires are about an inch and a half wide, and one and a quarter thick.   Wheels shod with them make no noise, and they greatly relieve concussion on pavements and rough roads ; they have lately been patronised by her Majesty.

and fibrous materials with rubber and sulphur and other matters, and vulcanizing them; these soles wear much longer than leather, and are of course waterproof, and always retain their elasticity. This material has also been applied to steam packing, pump-buckets, and other uses. We found, at a very early period of vulcanizing, that a substance resembling sponge could be readily produced of various qualities; it has been used as an elastic stuffing, and answers well for such purposes, but the dark colour renders it objectionable as a general substitute for sponge; this material can be moulded into any form; it has been proposed to employ it for billiard cushions, but as these require exact uniformity in their elasticity throughout, sponginess cannot be produced with sufficient accuracy; solid rubber cushions have long been used. I mention these to show that we were ready at once to supply whatever was asked for; my process of heating by steam was, in fact, the system which enabled us to appear so early and so successfully in the field with many of these articles; I have not extended the list of application here, as I intend to insert one at greater length hereafter.

Some time ago, the subject of preventing the destructive effect of shot upon iron ships was much discussed. With the view of ascertaining, as far as I could by mechanical means, how vulcanized rubber would operate in obviating this mischief, I proposed to Mr. James Nasmyth, the eminent

DOMESTIC ARTICLES.

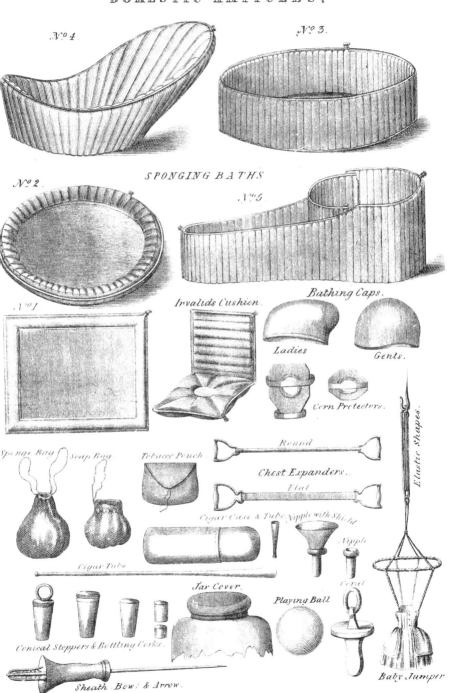

No 4.

No 3.

SPONGING BATHS.

No 2.

No 5.

No 1.

Invalids Cushion.

Bathing Caps.

Ladies.

Gents.

Corn Protectors.

Sponge Bag.

Soap Bag.

Tobacco Pouch.

Round

Chest Expanders.

Flat

Elastic Shapes.

Cigar Case & Tube.

Nipple with Shield.

Cigar Tube.

Nipple.

Jar Cover.

Coral.

Playing Ball.

Conical Stoppers & Bottling Corks.

Sheath Bow & Arrow.

Baby Jumper.

J Slater del & sculp

engineer, the trial of different thicknesses by the test of one of his steam hámmers, to which he readily assented, and very politely undertook to conduct the experiment himself at the works at Patricroft; I had three pieces of vulcanized rubber prepared, each a foot square; one an inch thick, one an inch and a half, and the other two inches. That the trial might be (although in a small degree) somewhat similar to the effect of a gun shot; I had an iron hemisphere cast six inches in diameter, thus forming the half of a six-inch ball; the steam hammer was first brought down upon the piece of vulcanized rubber one inch thick, upon which it produced no visible effect; the iron casting was then laid on the piece one and a half thick, with the flat side up; the hammer was brought down upon the casting, and broke it to pieces, but made no visible impression on the rubber. The two-inch piece was then laid on the anvil, and a round six-inch shot laid on it; the hammer was again brought down with tremendous force, which broke the shot into several pieces; on examining the fragments, we found that the shot had come into contact with the anvil, and was flattened slightly; we now examined the rubber, and found it had sustained no injury; on bending it sharply over the acute angle of the anvil, a slight incision could be perceived, three small cuts radiating from a centre. On another occasion, a pair of heavy rollers were set so as to produce a piece of lead passed between them a quarter of an inch thick.   With the rollers in this position, a piece of vulcanized rubber about seven inches square, and one inch and a half thick, was passed between them ; the rubber resumed its original form and size the moment it escaped from the grip of the rollers.

During the Crimean campaign I tried an experiment with a compound structure of metal and vulcanized rubber to serve as a kind of protection to the men engaged in the rifle pits, and had one made for trial by the military.   I gave it to some officers, but I heard nothing further of it.   I am inclined to think it should have had attention, as from trials on my model the results were promising.   I have had to convince several persons that rubber would not resist the force of a bullet ; they had fired at it with paper in front ; this made a hole in the paper, but none (as they said) in the rubber, they were unable to find the hole in the rubber ; because the shot had taken nothing out, and the hole closed up.   One gentleman had made an appointment to exhibit this feat before the Duke of York, and called on me for the rubber. I told him of his mistake, but he continued positive ; that he might be convinced, and saved from ridicule, I proposed a trial, he would have it as nearly as possible like shooting at a man.   I went into the garden, and had a breast of mutton fixed on the wall, and a piece of rubber attached to it.   I then took a rifle and sent a ball not only through the rubber and mutton, but half way through the wall

SPORTING ARTICLES.

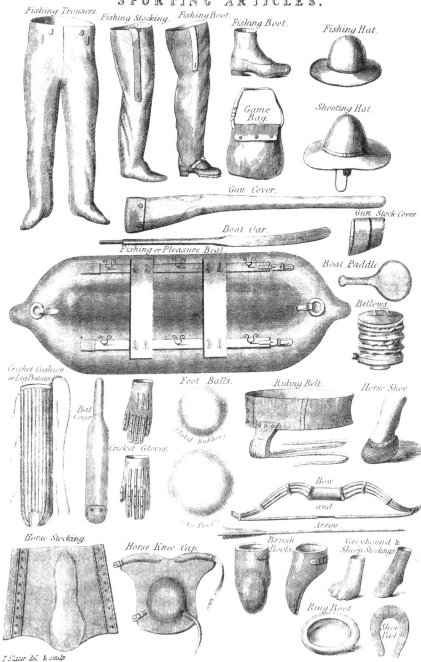

Fishing Trousers. Fishing Stocking. Fishing Boot. Fishing Boot. Fishing Hat.

Game Bag. Shooting Hat.

Gun Cover. Gun Stock Cover.

Boat Oar.

Fishing or Pleasure Boat. Boat Paddle.

Bellows.

Cricket Cushion, or Leg Protector. Foot Balls. Riding Belt. Horse Shoe.

Bat Cover. (Solid Rubber.)

Cricket Gloves.

Bow and Arrow.

(Air Proof.)

Horse Stocking. Horse Knee Cap. Brush Boots. Greyhound & Sheep Stockings.

Ring Boot. Shoe Pad.

T. Slater del. & sculp.

itself; the projector did not meet the Duke, who, I suppose, considered him a madman. Although the rubber of course was not vulcanized in those days, yet the result would have been the same.

In 1846, I obtained a patent for vulcanizing rubber in or upon moulds, plates, or forms, and retaining such articles under pressure or upon such forms during their vulcanization, by which means permanency of form is given to articles such as shoes, surgical bottles, valves, &c. Impressions from engraved plates so produced remain permanently raised. I may here mention that in this specification I described the hard or highly vulcanized rubber as one of the materials of which I formed my moulds. Although the means of producing this hard material was described in my first vulcanizing patent of 1843, this was, I think, the first time the actual application of it to any particular purpose had been published.

An almost endless variety of productions has sprung from the applications of this patent, as the most delicate chasings, tracings, and other ornamental decorations, as well as medallions, bas-reliefs, type for letter-press printing, surgical uses, toy-balls, both solid and hollow, moulding for a great number of mechanical purposes, some of which I have mentioned before, and others are almost daily occurring, some of them requiring a good deal of skill in the formation of the moulds, and experience in adapting and moulding the materials for the requirements of inventors and patentees, particularly in articles which were to be hollow, and yet to present on the external surfaces letters, figures, and embossed patterns. Provision was also made in this patent for the free escape of air or perspiration, in cases where needed, by perforations either plain or concealed by ornamenting the surrounding part. The means of carrying out all these in practice are described at length in the specification which will be found in the Appendix.

The hard vulcanized rubber has been applied to many useful purposes to which this patent has contributed. Combs, knife and other handles, ornamental panels for carriages and furniture, stop-cocks, tubing, pump-barrels, pistons and valves for use in chemical works, &c. &c.,—these are capable of being turned in the lathe, and to have screws cut on them in the same manner as is practised with wood, ivory, or metal. I have also had some flutes made of it, the colour is a jet black, and it polishes like ebony; the notes or sounds are equal to the best flutes, whilst they are said to be produced with greater ease by the performer. I furnished the material to the flute maker without instruction, and he made it in his ordinary practice.

This patent for moulding and giving permanency to forms by vulcanizing has, in fact, tended to bring the hard "changed" rubber prominently into view; and an endless variety of articles have been and may be made by means of this patent, besides the uses to which it is being applied as the raw material, from which to manufacture various articles. We have supplied it by the ton for the use of comb-makers, who like it, not only because it makes a good saleable article, but because they can have it in large sheets of the thickness they require, and make much less waste than when using such small pieces as are produced in horn, tortoise-shell, &c. There is no other limit to the extent of its employment, except that which arises in point of utility, appearance, or economy in competition with other materials. The turner, the engraver, the comb-maker, and most other artists and mechanics, have only to apply their ordinary means, tools, and skill, as to wood, ivory, metal, and other substances. It is also a fair substitute for whalebone and walking-sticks, and also for more delicate articles, as bracelets, gold and silver mountings, pens, and penholders, picture-frames; and one may go from these to the contrary extreme, and if it were economical, or in any way advantageous to do so, it would make good houses, ships, waggons, and carts, and almost everything where wood is now employed, which I only mention to show the universality of its application, and in general, by the ordinary means practised in the different departments of Art. By this process rubber can in the same piece be rendered hard in one part, and elastic in another, or brought to any intermediate state, that is, flexible, but inelastic like the firmer kinds of leather. There seems to be little room for invention in its application to these objects;

this moulding patent and ordinary means are sufficient.

We exhibited embossings of this hard horny material in our stall in the Great Exhibition of 1851, and in the Crystal Palace, among our other manufactures. In the hard vulcanized rubber, however, we have confined ourselves chiefly to slabs, bars, and sheets for manufacturing purposes, and to articles of known utility, and for purposes in which its peculiar characteristics rendered it of value, and in some cases, where heretofore no material had been found to answer so well, and to others where its adoption superseded, with more or less advantage, such other matters as had previously been used. Although this has been our ordinary business course, yet we have, nevertheless, kept an assortment at our warehouses both in London and Manchester, including some beautiful specimens of artistic skill, both in design and execution, also comprising mouldings of the human hand, portraits, tablets, cylinders, articles of furniture, stick and umbrella handles, door knobs, printers' type, embossed lettering, inlaying, metallic ornamental specimens, marquetry, buhl work, and many others ; some of these we have no intention of becoming manufacturers of, but keep them simply as samples to show to those whose business it is to do so, not only that we make and sell this material, but also to exhibit the nature of it, and the facility with which it can be worked.   These were made for the purpose of introducing this new substance to the notice of the

public, as cabinet-makers, bookbinders, carvers, op-
ticians, jewellers, turners, or musical instrument-
makers, and others who would of their own more
intimate knowledge of their different branches and
pursuits of course suggest—indeed, have suggested,
and will no doubt suggest, — the adoption of this
material according as people in trade see their way
to an advantageous or economical use of it as compared with other substances heretofore employed
for their respective purposes.

In the spring of 1846 a patent was taken out by
Mr. Alex. Parkes for a process by which he produced in rubber nearly the same effects or properties possessed by vulcanized rubber, and gave it the
name of "converted rubber." His process is an
elegant and simple one, and consists in immersing
the rubber in a solution of the chloride of sulphur
in bisulphurate of carbon, or pure coal naphtha cold,
no heat being required; a thin sheet of rubber
is by this means "converted" in a minute or two,
and when dry is found to have acquired the properties of insolubility at ordinary temperatures, and
to be insensible to cold. The process is capable
also of producing the horny state, similar to hard
vulcanizing.

The results effected by this process are as mys-
terious as those by vulcanization; they could not
have been anticipated, they cannot be accounted
for, and they prove moreover that heat is not absolutely necessary to produce this obscure change in
the condition of rubber.

The mention of this patent would be hardly consistent with a "personal narrative," but for the reason that I assisted Mr. Parkes in his experiments and specification, and that this patent ultimately came into our possession, and also that in the autumn of the same year I obtained in conjunction with Mr. Brockedon a patent for various applications of Mr. Parkes's process. This patent embraced the manufacture both of rubber and gutta percha, either separately or combined, in all proportions. These substances being very similar in character, they united readily in the masticator, and are dissolved by the same menstrua, and capable of being spread and moulded in the manner and by the means already described in the specifications before described for rubber. Articles formed of either of these substances, or compounds of the two, and also with fibrous substances and colouring matter, are submitted to the converting process of Mr. Parkes, and are thereby rendered insoluble in essential oils, unaffected by oil or grease, and not subject to change by exposure to variable temperatures. These articles are also by the converting process rendered as hard, or if necessary harder than ivory, and capable of being wrought with tools and highly polished, in the same manner as before described in respect to hard vulcanized rubber. The specification will be found in the Appendix.

During the course of these experiments and the adaptation of vulcanized rubber to so many purposes, I am desirous of taking this opportunity of recording the faithful and able services of the manager of our works, Mr. A. B. Woodcock; his energy and zeal in executing the duties of his position entitles him to honourable mention in my narrative as a memorial of the active part he has taken in the process of this manufacture.

This converting process can only be conveniently employed upon rubber when of a thin form or substance, as the chloride acts rapidly upon it; and if the rubber is thick, the surface becomes hard and brittle before the interior is affected. We have found this invention in many cases a useful auxiliary to vulcanization, and in the absence of the latter it would have been of considerable value, as by it nearly everything can be done which could be effected by vulcanizing in a hot stove or by any dry heat.

The process of Mr. Parkes enables us to give to vulcanized articles colours of every tint, and a delicately smooth surface. These converted surfaces also print well, and the most delicate impressions from copper-plate engravings are produced upon them. Gutta percha and compounds of this substance with rubber are equally susceptible of improvements in the same way.

Observing that the elastic web imported from France was ornamented with designs in various colours effected by expensive and peculiar modes of weaving them, the thought struck me that patterns migh be printed on them, and on trial

I found this could easily be done, and some novel results ensued. The web could be printed of any given pattern in its contracted state ; and then, on extending it, all the objects in the pattern were extended in due proportions until it became like a dissolving view almost lost to the sight. On allowing it to contract, the lost pattern and colours were restored in all their freshness ; if, on the contrary, the printing was done whilst the web was in its extended state, a curious concentration or condensation of the pattern was produced. I obtained a patent for this improvement during the year 1847. The specification is in the Appendix.

In the early part of this year we had to lament the loss of another of my partners, Mr. Joseph Birley, whose decease took place on the 24th of January, 1847. He gave his hearty concurrence to the efforts of his partners in the improvements of the rubber manufactures, and took pleasure in watching their progress, but his attention to other business occupations prevented him from doing more than affording us the benefit of his sound judgment and great experience when in consultation. His memory is cherished amongst us, and his loss has left a blank which we still deplore.

Many attempts have been made to discover some means of dissolving or otherwise using up the waste of vulcanized rubber, and in the course of this year I made many experiments with this view, assisted by a gentleman who had a more

intimate acquaintance with chemistry than I possessed; the best result, however, that we arrived
at, was its solution by means of oil of turpentine
at a very high temperature, or by coal naphtha,
the boiling point of which being much too low,
we were obliged to resort to very strong closed
vessels, and then, if the rubber had been vulcanized
to a hard or horny state, the process was very slow.
The product is not very tractable, and requires a
good deal of trituration.

A patent was obtained for these modes of operating on vulcanized waste during this year; this
patent also included an improvement in moulds
for rubber, which consisted in forming them of
materials easily dissolved or melted out, such as
compounds of gum or glue and whiting, or the
fusible metal known as Darey's alloy, which melts
in boiling water.   This specification will be found
in the Appendix.

In the early part of this year (1847) we were
apprised of the sale of some American over-shoes,
and on procuring some of them we found they
were made of vulcanized rubber.   Notice was
therefore given to the parties that they were
infringing my patent; this was denied, and we
therefore gave notice that an action would be commenced against them, which was done.   The shoes
were of the manufacture of the Hayward Rubber
Company in Connecticut, and one of the firm
came to this country and made an arrangement
with us.

We had not done much in vulcanized over-
shoes, nor did we attach any great importance to
that business, and I must frankly own I believe I
thought less of it than my partners, remembering
the fuss made about the free escape of insensible
perspiration, and the unpleasantness in this respect I had always found (myself) in wearing
rubber shoes; my judgment has been proved by
the public to be at fault, and vulcanized rubber
shoes have had an immense sale.   We granted the
Hayward Rubber Company an exclusive license
for the importation of American over-shoes of
their own manufacture on moderate terms, which
license exists at the present time.

Our over-shoes of pure vulcanized rubber were
very neatly made, of good forms, and superior to
the American in lightness and elasticity, but they
had very little shine or gloss upon them, and although they were liked by many, yet the polished
ones were generally preferred.   These contained a
good deal of lead and other matters, and an oily
polish or composition or varnish added by the
shoe manufacturers, for which each manufacturer
had his own secret recipe.   Our over-shoes, being
of pure rubber, were vulcanized in a steam boiler,
by which means the utmost elasticity was attained;
but this mode is obviously incompatible with a
glossy surface.

The year 1851 brought with it the memorable
Crystal Palace and the "Great Exhibition of the
Works of the Industry of all Nations," and we

TRAVELLING ARTICLES.

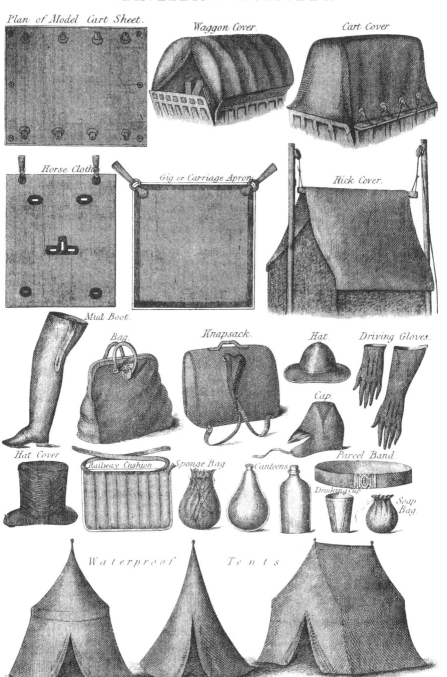

Plan of Model Cart Sheet.

Waggon Cover.

Cart Cover.

Horse Cloth.

Gig or Carriage Apron.

Rick Cover.

Mud Boot.

Bag.

Knapsack.

Hat.

Driving Gloves.

Cap.

Hat Cover.

Railway Cushion.

Sponge Bag.

Canteens.

Parcel Band.

Drinking Cup.

Soap Bag.

Waterproof   Tents

Slater del & sculp.

were not slow in availing ourselves of this opportunity of exhibiting such a general collection of
rubber manufactures as the world had never
before seen; comprising specimens of almost every
article to which the substance had been applied,
whether adhesive or unadhesive, vulcanized or unvulcanized, possessing elongating elasticity, or rendered rigid by hard vulcanizing, plain, coloured,
printed, embossed, moulded portraits, medallions,
tablets, stick and umbrella handles, mechanical
applications, toys, and various other things made
entirely of rubber, and ordinary and coloured solutions were also there, to which must be added
some beautiful specimens of rubber produced by
the converting process of Mr. Alexander Parkes.
Of course we had also all the well-known Macintosh articles, such as cloaks, capes, of double
and single textures, air-beds, pillows, cushions, life-
preservers, model pontoons, diving dresses, gas-
bags, &c. &c.

We had the pleasure of witnessing the notice
taken of our stall by Her Majesty and Prince
Albert ; the latter of whom took with him a tablet
of vulcanized rubber, on which a few lines from
Cowper were embossed ; they were so appropriate
to the national occasion that I am tempted to
insert them ; they were surrounded with the
national arms, and with the rose, shamrock, and
thistle, as an ornamental border : —

COMMERCE.

... " The band of commerce was design'd
T' associate all the branches of mankind ;
And if a boundless plenty be the robe,
Trade is the golden girdle of the globe.
Wise to promote whatever end he means,
God opens fruitful Nature's various scenes ;
Each climate needs what other climes produce,
And offers something to the general use ;
No land but listens to the common call,
And in return receives supplies from all.
This genial intercourse and mutual aid
Cheers what were else an universal shade,
Calls Nature from her ivy-mantled den,
And softens human rock-work into men.
Ingenious Art with her expressive face
Steps forth to fashion and refine the race,—
Not only fills Necessity's demand,
But overcharges her capacious hand ;
Capricious taste itself can crave no more
Than she supplies from her abounding store :
She strikes out all that luxury can ask,
And gains new vigour at her endless task.
Hers is the spacious arch, the shapely spire,
The painter's pencil, and the poet's lyre,
From her the canvass borrows light and shade,
And verse more lasting, hues that never fade."

COWPER.

May 1st, 1851.

I give an extract from the Report of the Jurors,
who awarded to us in flattering terms the Council
medal : —

" COUNCIL MEDAL."

" Charles Macintosh and Co., 73. Aldermanbury
(now 3. Cannon Street, West).—The firm of Charles

Macintosh and Co. comprises the names of the men who in Europe have made the most useful discoveries in the art of applying caoutchouc to the most varied uses : — the late Mr. Macintosh, who gave his own name to the waterproof garments, and Mr. Thomas Hancock, whose share of merit in the discovery of the vulcanization of india-rubber we have already mentioned. In going through the collection of articles exhibited by this firm, the importance of the uses to which the substance is capable of being applied, especially since the discovery of the process of vulcanization, can be readily appreciated.

" The kinds of fabrics with which the garments called Macintoshes are manufactured have always remained the same, but the garments themselves have acquired more lightness and less smell, and the substitution of vulcanized for common caoutchouc insures to them at the present day a permanent suppleness.

" The other services which these fabrics are called upon to perform have been greatly multiplied. Their price having become less, they are capable of being applied in lieu of tarpaulings for covering waggons, carriages, &c. The property which they possess of serving to contain water, and which had at first been made available for a well-known therapeutic use, has allowed of their being made into portable baths, which can be rolled up like an ordinary cloth when not in use. The shoes exhibited by Messrs. Macintosh and Co. are made

with much care, and with a degree of elegance,
which shows that in Europe these articles are but
little used except by the more opulent classes.

" It is not only in the making of shoes that india-
rubber has been called in to supersede leather;
the articles exhibited by Messrs. Macintosh and
Co. show the use that can be made of it to form
pistons of pumps, and how conical valves of
india-rubber can be advantageously substituted
for leather or metal ones. Sheets of caoutchouc of
different colours, either smoothed or worked in
relief, are brought in to supersede moulded ornaments in the manufacture of furniture, of ottomans,
and in the binding of books.

" The use of vulcanized india-rubber to form the
piston-valves in steam-engines on the screw principle has greatly contributed to the employment
of these novel motive powers, which are destined
in some degree to effect a change in navigation, by
allowing steam to come in solely as an auxiliary to
the wind. The exhibition of Messrs. Macintosh
and Co. comprises a valve of this description,
which after six months' use has undergone so
little alteration that it may be foreseen that these
articles possess an almost unlimited durability.
The rendering available the impermeability of their
fabrics to gas and air has likewise been extended.
To the air-cushions which have been long used are
now added the air-mattresses, so well adapted as
beds for travellers and invalids, boats inflated with
air at once portable, and incapable of sinking, and

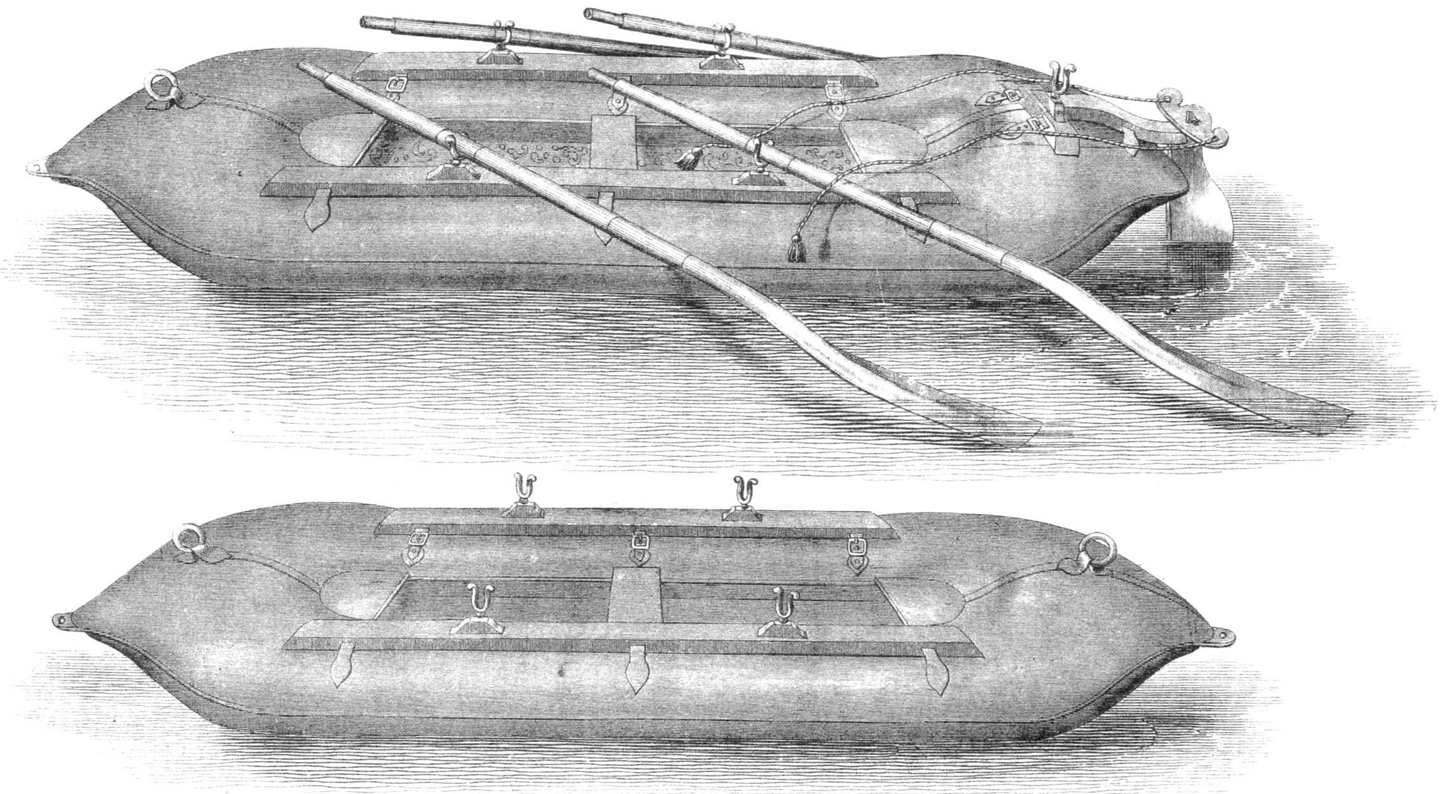

which for life-boat uses and in inland voyages are capable of rendering great service. The collection of Messrs. Macintosh presents some specimens of this kind of great interest. Lieutenant Halkett of the Royal Navy, by making them in several closed compartments, in such a manner that the stuff being pierced through at one point cannot lead as a necessary consequence to the sinking of the boat, has rendered more certain the services which these machines are called upon to perform. The applications of the elasticity of caoutchouc have also been greatly increased. The wheels of carriages have been surrounded by it in such a manner as to prevent the disagreeable noise which they make upon the pavement. Rollers for inking printing-types and lithographic stones have been made from it; it is employed for the making of cushions for billiard-tables, and to supersede the use of sacking cords in bedsteads. Advantage has been taken of the elasticity of an india-rubber band, which has a tendency to return to its primitive length when the action of opening a door has elongated it, for the purpose of forming door-springs, the use of which is beginning to spread widely.

" The substitution of vulcanized india-rubber for metallic springs in the buffers of locomotive engines is one of great service. The masses of vulcanised india-rubber deaden the shock with an ease which may cause this employment of caout-chouc to be considered as one of the most useful to which it has been applied up to the present day. A certain number of the novel applications are due to Messrs. Charles Macintosh and Co. The ability which they have displayed in the manufacture of caoutchouc has afforded to other inventors of these new applications the means of putting their ideas into practice. Without the discovery of vulcanized india-rubber, they could, moreover, never have been carried out. The jury, therefore, in order to recompense the considerable services rendered in the employment of caoutchouc by Messrs. Macintosh, Hancock, and their other partners, have awarded a council medal to the firm of Charles Macintosh and Co."

During the spring of 1852, Professor Brande delivered a course of lectures on organic chemistry before the members of the Royal Institution, and he has kindly given me permission to make use of any part of them.

As these lectures were delivered twenty-six years later than the paper of Professor Faraday before mentioned, they contain new matter on the chemical part of my subject, and I have thought that information from such a source would not prove unacceptable to some of my readers, although not quite consistent with a personal narrative: —

"Caoutchouc is sometimes called *gum-elastic*, but it has none of the characters of a gum, being perfectly insoluble in water ; nor is it a resin, for it is insoluble in alcohol; but ether, which has been washed, so as to deprive it of alcohol, dissolves it, with the exception of impurities, and leaves it, on evaporation, in its perfectly elastic state; it also dissolves in chloroform, and in sulphuret of carbon, and in some of the essential oils, but its most common and useful solvents are rectified naphtha and oil of turpentine. This substance came first into notice about the beginning of the last century, moulded into the shape of bottles and animals. It was sold as high as a guinea the ounce, and used for rubbing out pencil marks; but scarcely anything was known of its history, except that it came from America, till De la Condamine sent an account of it to the French Academy in 1736, describing it as the inspissated juice of a tree, called by the natives *hevee*. In the year 1751, Frisnan discovered the same tree in Cayenne. It is now known to be the produce of several trees growing in South America and in the East Indies. It exudes from fissures and wounds of the trees, in the form of a milky juice, which, exposed to air, soon forms the elastic deposit. It is very inflammable, burning with a dense smoky flame. It burns brilliantly in oxygen gas, but even then throws off abundant flakes of unburnt carbon. When subjected to distillation in close vessels, it yields an oily liquid which has a singularly disagreeable and penetrating greasy smell, and appears to be a mixture of several hydrocarbons, one of which predominates, and has been called caoutchine.

" The first person who applied solutions of india-rubber largely in manufactures was Mr. Charles Macintosh of Glasgow; and the waterproof tissues which he made were well known by his name.

" About the year 1843, Mr. Thomas Hancock, of London, made a very curious and important discovery in respect to caoutchouc, which has given an entirely new aspect to almost all the manufactures dependent upon it, and by which its properties are so modified and improved in respect to the great majority of its uses as to constitute it, as it were, a new material. In this altered condition, it is now known under the name of *vulcanized india-rubber*, and, as the whole process is a very remarkable one, I will endeavour to show you the various steps by which it is effected. The abundance of articles of all kinds, which are upon the table before me, will speak for themselves as to the utility and importance of this remarkable invention. I am indebted to Mr. Hancock, and generally to the firm of Messrs. Macintosh and Company of Manchester, for the specimens before you. I have visited their factory and examined the numerous operations carried on there, in all their particulars, so that, although my time obliges me to be very brief in the details I now wish to illustrate, I shall be happy, after lecture, to give any further information upon them which may be required.

" India-rubber, as imported, generally includes mixed impurities of different kinds, which are got rid of by tearing it up into shreds, in a stream of water; it is then transferred to a machine called a *masticator*, where it is severely pinched and kneaded till it forms a kind of mass or *magma*, which may be compressed into blocks, cut into sheets or threads, or rolled out into layers as thick or thin as we please; but all these, and similar operations, leave the rubber with its original objectionable qualities, namely, its clamminess, its tendency to stiffen and harden, and consequently to lose its elasticity by cold; to be readily softened and decomposed by heat; and to be rendered soft and, as it were, rotten by grease and oils. As to the clammy adhesiveness of india-rubber, Mr. Hancock gets rid of that by thoroughly and completely blending it with a quantity of finely powdered French chalk, or silicate of magnesia, the well-known soft, silky feel of which is thus imparted to the rubber.

"The other objectionable properties are subdued by the extraordinary influence of *sulphur*; and it is to its operation that I wish particularly to direct your attention. Sulphur is, in any convenient way, in the first instance, *mixed* or *blended* with the rubber; as by solution, or rolling, or in the masticator, or dipping into melted sulphur. I have here a mass of india-rubber, which contains about ten per cent. of flour of sulphur, well mixed or blended with it; but as yet the common properties of the rubber are unaffected. If, however, we now expose this mixture of rubber and sulphur for a due time to a high temperature, which may vary with circumstances, a change ensues. I here heat it to about 300°, and find the characters of the rubber gradually alter. Ultimately it will be *changed* or *vulcanized ;* that is, it will have ac-quired the new characters and properties to which I have briefly alluded. Its elasticity is not only increased, but is now not affected by change of temperature; that is, it is not in the least hardened or altered by any degree of cold; it retains its perfect elasticity at the lowest temperatures ; and what is perhaps even still more remarkable, when heated, it does not fuse and become clammy and viscid, but remains unchanged at all temperatures short of its absolute decomposition; ether, oil of turpentine, naphtha, and its other solvents, now barely affect it ; oil and grease no longer penetrate and soften it. Such are the *new properties* conferred upon the rubber by the joint operation of sulphur and heat; it is now said to be *vulcanized.*   But the most curious part of the story remains to be told, and arises out of this question, namely, What quantity of the sulphur is required to be thus com-bined with the rubber for the purpose of *changing* it ?   Certainly not more than one or two per cent.; for if by any of the solvents of sulphur, as for in-stance by the alkalies, I remove all excess of sulphur, I in no way affect the new properties; but yet fur-ther, if I take steps still to remove more of the sulphur, and ultimately leave the rubber apparently quite free from it, I still find that it retains all those peculiarities which I have enumerated as characterising vulcanization. May we not therefore conclude that, under the influence of sulphur and heat, the rubber acquires its new and distinct properties, not by actual chemical combination with a minute portion of sulphur, but by the assumption of a new *molecular condition*; that, like phosphorus, it has assumed an *allotropic state*. In short, these allotropic conditions of bodies are daily becoming more worthy of remark and inquiry; thus, it is not unlikely that *steel* even may be an *allotropic condition of iron*, produced under the influence of carbon upon iron at high temperatures, for the quantity of carbon in the finest steel does not reach *one per cent.*; and there are cases in which the metal is *nearly* pure to all chemical tests, and yet retains the leading characters belonging to steel."

Sulphur has been mentioned all along as the substance to which vulcanization is undoubtedly due, yet it may be as well to observe, that it is not meant that this substance must be always employed in its primitive state; as the forms in which it may be employed are various, so also are the names given to the same results by different parties, — such as "galvanised," "mineralised," "metalised," "thionised," &c., by which the process of vulcanization is really meant.

I have observed that it has been thought, in some quarters, that rubber, before the discovery of vulcanization, had scarcely been brought into any extensive use; what has already been said will serve to correct this error, and the fact that we were frequently using from three to four tons a week, as early as 1837 and 1838, is a further proof of its usefulness in its unvulcanized state ; at this period, I believe, there had been very little done in anv other country except France, although here and there some beginnings were being made or attempted. Many of the articles we have been making for upwards of thirty years are still preferred of unvulcanized rubber, and they would have continued to be made had vulcanization never been discovered; but nearly all the articles made of the old material are greatly improved by substituting the new. The great benefit, however, derived from the discovery of vulcanization consists chiefly in its application to new purposes, for which it was before totally unfit, and could not be used. Rubber that has undergone any process of manufacture is weakened in its elastic property, which by this process is at least restored, if not considerably increased, and, what is of equal or greater importance, this elasticity is unaltered by cold, and therefore always opposing the same uniform resistance to concussion, as in railway buffers, and carriage and other springs. Its extensile elasticity and resilient action becomes permanent, as in elastic web, shoes and other articles ; its endurance of heat, and retention of its elasticity at temperatures at which it would formerly have become decomposed, as in packings of various kinds for steam-engines; and particularly the pump-valves of ocean screw steamers, — where

MECHANICAL PURPOSES.

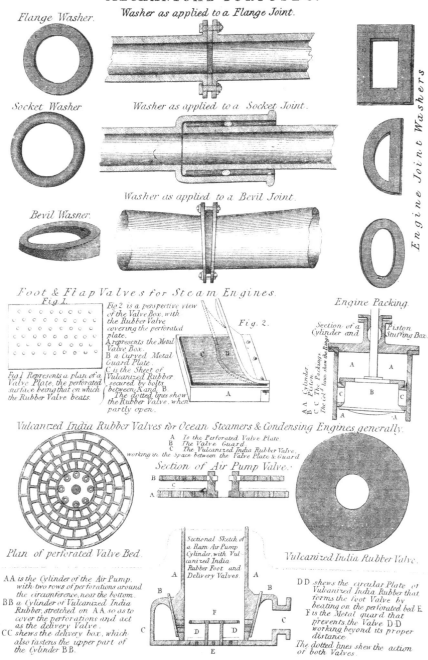

Flange Washer.

Washer as applied to a Flange Joint.

Socket Washer.

Washer as applied to a Socket Joint.

Engine Joint Washers

Bevil Washer.

Washer as applied to a Bevil Joint.

Foot & Flap Valves for Steam Engines.

Fig 1.

Fig 2 is a perspective view of the Valve Box, with the Rubber Valve covering the perforated plate.
A represents the Metal Valve Box.
B a Curved Metal Guard Plate.
C is the Sheet of Vulcanized Rubber secured by bolts between A and B.
The dotted lines shew the Rubber Valve when partly open.

Fig. 2.

Fig 1 Represents a plan of a Valve Plate, the perforated surface being that on which the Rubber Valve beats.

Engine Packing.

Section of a Cylinder and Piston Stuffing Box.

A A Cylinder.
B Piston.
C C The Packings.
The dotted lines shew the Rings.

Vulcanized India Rubber Valves for Ocean Steamers & Condensing Engines generally.

A   Is the Perforated Valve Plate.
B   The Valve Guard.
C   The Vulcanized India Rubber Valve, working in the space between the Valve Plate & Guard.

Section of Air Pump Valve.

Plan of perforated Valve Bed.

Sectional Sketch of a Ram Air Pump Cylinder, with Vulcanized India Rubber Foot and Delivery Valves.

Vulcanized India Rubber Valve.

A A is the Cylinder of the Air Pump, with two rows of perforations around the circumference, near the bottom.
B B a Cylinder of Vulcanized India Rubber, stretched on A A. so as to cover the perforations and act as the delivery Valve.
C C shews the delivery box, which also fastens the upper part of the Cylinder B B.

D D shews the circular Plate of Vulcanized India Rubber that forms the foot Valve by beating on the perforated bed E.
F is the Metal guard that prevents the Valve D D working beyond its proper distance.
The dotted lines shew the action of both Valves.

I. Slater. del. & sculp.

its long endurance, exposed to perpetual concussion and heat, is most extraordinary, and it is questionable whether any other substance could be substituted for it, or, at least, with the same advantages.

I have a letter now before me from Messrs. T. B. Palmer and Co. of Jarrow, Newcastle-upon-Tyne, in which they state that they fitted a screw steamer built by them, the "John Bowes," with our vulcanized rubber valves, which have been in use four years, during which time the vessel ran upwards of 100,000 miles, the speed of the engines sometimes upwards of 100 strokes a minute; usual speed, 75 strokes a minute.

I will not enumerate more, but simply state that engineers, machinists, surgeons, chemists, and others find in this new state of the rubber a fitness for their purposes which is daily more and more developed.

The vulcanized rubber thread has lately been introduced into the Jacquard loom, by Messrs. Bonnet and Co., Manchester. The thread is used to supersede the use of weights; the number of these sometimes amounts to from two to three thousand in one loom.

Although all this is due to the altered state of the material, yet it may easily be imagined that no small amount of thought and skill have been brought to bear upon these applications, to provide apparatus, and to devise means to carry into execution the requirements of the various parties who have discovered its fitness for their respective purposes, and the applications of this new substance to our general business in water-proofing, moulding, &c. &c.

I have mentioned that in 1847 we had granted an exclusive license to the Hayward Rubber Company in America, for the importation of vulcanized rubber over-shoes of their own manufacture. These over-shoes sold slowly at first, as they were rather heavy and clumsy, but in America they speedily became popular, and many were soon engaged in their manufacture, and consequently a rather rapid improvement was made in the article, and our licensees were amongst the foremost in the race. The shoes eventually appeared with a better polish, and their forms and general appearance became such as to invite the attention of shoe-dealers, and through them the attention of the public. Their usefulness also began to be appreciated, and in the course of time they came to be very generally sold. Under these circumstances the other manufacturers in America, as well as the shoe-dealers here, became jealous of the Hayward Rubber Company and their exclusive license.

However much any of these parties may have been affected by the license, we were not to blame, for I doubt whether, at the early period at which the license was granted, there existed any other considerable manufacturers of these articles in America, or at all events I do not remember that we had any knowledge of any such existing, the Hayward Rubber Company being the only parties that applied for a license. The shoes as then made were, as before observed, very inferior in quality to those sold afterwards, and the trade in them at that time quite in its infancy.

Whatever might have been the feeling amongst rival manufacturers elsewhere, it is not easy to imagine why the shoe-dealers in this country should not have been content to have taken their shoes from our licensees (who paid only a very moderate royalty); they would then have all been served on equal terms, and the competition which followed, arising from an opposite course, would not have occurred, nor an excellent and profitable trade spoiled by it, as it since has been. Unfortunately, the dealers decided on disputing the validity of my patent, and subscribed a purse among them to defray the expenses. I shall not trouble myself or my readers by entering at any length upon the subject of litigation; suffice it to say that an action was tried in the Court of Common Pleas on the 20th and 21st June, 1851, and was concluded by the jury returning a verdict for me on all the counts, and the judge certified that the validity of the patent came into question in the action.

I have now brought my narrative up to the latter end of the year 1854, and to the eleventh year of my patent. It may therefore be readily imagined that I felt pretty well assured that my patent would now be allowed to run out its remaining days in peace; and, indeed, both myself, my partners, and our esteemed solicitor, Mr. Henry Karslake, who had conducted our various law pro-ceedings with great ability, and whose earnest zeal and indefatigable exertion never tired, and will never be erased from a grateful remembrance, were all of us weary of a long course of litigation (which had been kept up for seven years), and required the repose that now seemed to have been secured.

It has pleased the Almighty that I should out-live all my original partners, and during the long course in which we have prosecuted our manu-factures and business, a kindness of feeling and a cordiality has existed which renders the task of recording another loss — though of a later partner — painful to me, as the same feeling still animated us all. I now allude to Mr. Brockedon, who died on the 29th of August, 1854. He had for some years paid considerable attention to applications of rubber to various purposes, and was well informed in all that related to its manufacture, and it is to him that I owe the term "vulcanization," by which name rubber in its new state is now universally called. His literary works, his extensive informa-tion, his scientific acquirements, and his bland and gentlemanly bearing, are too well known to need more than a passing memento in my simple narra-tive.

It has been said that "security is dangerous when men will not believe any bees to be in the hive until they have a sharp sense of their stings," and so I found it; for on my birth-day, the 8th of

May, 1855, in the sixty-ninth year of my age, I had a writ of *scire facias* issued against my long tried patent; and on the 7th of July, exactly twelve years after I applied for it, this writ came on for trial. I will not dwell upon it. The prosecutors did not promote delay, and after two day's trial in the Court of Queen's Bench, before Lord Chief Justice Campbell and a special jury, a verdict was returned in my favour, being now the second clear verdict which I had obtained. This process of *scire facias* so severely tests a patent that most that have been exposed to its sifting character have been lost whilst passing through the ordeal. It appears since the trial that if the parties who promoted this proceeding had clearly understood our respective positions it would never have been taken, and they subsequently agreed to terms, for the use of the patent, for the purposes to which they were desirous of applying it. The other infringers against whom proceedings had been taken submitted to terms, and the trade of vulcanizing rubber is now carried on in this country by ourselves and our various licensees.

If I could have known in 1843 that the laws of the land would have undergone such a sweeping change as that I should be called upon thereafter to give evidence upon oath of all the minute particulars passing in my laboratory during my experiments whilst making this discovery, I might have taken care to have had other corroborating testimony; but I was not possessed of prescience,

and had no remedy for the want of it. Supposing
I had employed an assistant, there was no provi-
sional specification to protect me in those days, and.
I might have been betrayed, especially when it is
considered how extremely uncertain are the
opinions of the judges on patent law, and particularly as to what constitutes a publication fatal
to a patent. I ought to explain that before the commencement of these two last trials the new law of
evidence had come into operation ; I could therefore
be examined on my own behalf, and this doubtless
in many cases might be a great privilege, and was
so in some measure to me, but it was nevertheless
in such a case as mine an anxious and arduous task,
and one on which I would not have entered, but that
I might have been called on the other side against
myself. In ordinary cases all would in general be
easy enough ; but the reader can have no idea, nor
is it possible for me to convey by language the
vast difficulty of being prepared or such an ordeal
as I had to undergo, nor could the legislature,
when passing the new law, ever have contemplated
such a case. I will just mention some of the
allegations which were preferred against the
validity of my patent: — that I was not the first
inventor in this realm; that I had not invented
that which I claimed; that if I had invented it,
I had not invented it at the time I applied for my
patent; that there had been a publication of the
invention by the exhibition of articles made in the
same way, &c. &c.

Now, reader, you may judge of the hardship of my case. How could I *prove* that I was the first inventor ? How could I prove that I had made the invention at the time I applied for my patent ? I could assert it, but how could I prove it ? Twelve years had elapsed, I had grown old ; and my memory was failing. I worked alone ; not a human being entered my laboratory during the whole time of making those tedious experiments I have mentioned ; I could, therefore, call no witness, I had none. I kept no record, not having the least idea that I should ever have to give any account whatever of these proceedings : how was I to call to mind the minute circumstances ? the order ? the tages of progress — the moments of favourable appearances, and the substances or circumstances that attended or occasioned them ? And yet this was pressed upon me with all the subtility of one of the most astute and able counsel that could be procured, sifting, catching at every word — perverting or altering my language — urging me to state the minutest particulars during my researches and experiments, in the endeavour to bring to light a discovery, acknowledged — be it remembered — when made, to be one of the most abstruse and obscure in science. Under this pressure, I could only have recourse to my failing memory, as to what I did myself; and to such circumstances as I could bring to mind or obtain, corroborative more or less of the facts. I was enabled to grapple with all these proceedings in a manner that surprised

myself; at the suggestion of my able and indefatigable solicitor, I ransacked all my drawers,
boxes, and receptacles, and amongst my experimental scraps, found plenty of very early ones that
were vulcanized; but mostly without dates. There
was fortunately, however, one dated the 4th of
June, and several in August, one of which was
hard vulcanised.   Of these dates, I could be quite
positive, although they were nearly obliterated;
and the scraps were ragged, insignificant little
things.   Then my servant could prove that ice
was taken in daily for my private use, from the ice-
cart that passed my gate ; she could also prove
that when my laboratory was pulled down in
March, 1843, I used the kitchen oven during the
summer till it was rebuilt; I also found the bills
for sulphur (with regular dates) procured in the
village at the time.   These not only assisted my
memory, but were in themselves all substantial
evidence, and when listened to with attention by
an intelligent jury, they readily found a verdict in
my favour, and came to me after the trial, and said
they had never heard evidence given with more
clearness and satisfaction to themselves.

This is only a very faint outline of what affected
me personally in these contests. I have said nothing
of the intrinsic evidence of science necessary to be
obtained ; nor the attending to purchases to prove
infringements : often purposely made difficult of
proof by the devices of the infringers and their
advisers, procuring analysis by scientific witnesses;

endless time at the office of our solicitor, attending consultations of counsel, and answering innumerable questions as to details of manufacture, &c.

Besides these law suits, there were injunctions to be moved for; affidavits to be prepared and sworn both by myself and the scientific gentlemen and others; attendance before masters in Chancery; and attendance at the courts on the different motions, judgments, &c. &c.

Though called upon to do so, we have never interfered in the disputes on this question abroad, nor have we made any attempt to monopolise any portion of the rubber trade in America, or any other foreign country.

I am greatly indebted to Dr. Lindley, who at my request has been so good as to furnish me with the following notice of the plants that yield the best caoutchouc.

" Of the plants that yield commercial caoutchouc, the most important are the following : —

" I. *Siphonia elastica.* —This, which yields the caoutchouc of Para, is a tree inhabiting dense forests on the banks of the river Amazon, and several of its tributaries, where it is called the Seringue.   The chief district from which its caoutchouc is obtained is, according to Wallace, the country between Para and the Xingui river. Aublet speaks of it as also occurring in the forests of French Guiana, where it is called Siringa by the Garipon Indians, Hevé by the natives of Esmeraldas, and Caoutchouc by the Mainas.

" According to this author, the trees are from fifty to sixty feet high, and from two to two feet and a half in diameter.   The bark is greyish, and by no means thick; the wood is white and light. The leaves each consist of three or more blunt leaflets attached by a joint to a long slender footstalk, and having an oblong form narrowing to the base ; they are green above, but ash-coloured on the under side.   The flowers are small, greenish, in long, loose bunches ; the fruit is about as big as a walnut, with a rind that separates of itself, and a hard bony shell splitting with elasticity into half a dozen pieces.   In each cavity of the fruit are found from one to three seeds, about as large as

Tho.ˢ Hancock Jun.ᵗ Delᵗ.

Alfred S Potter, Lith.

SIPHONIA ELASTICA.

Drawn from a Plant in the Conservatory of the Royal Botanical Gardens at Kew.

Tho.ᵈ Hancock, Jun.ᵗ, Delt.

Alfred. S. Petter, Lith.

HANCORNIA SPECIOSA.

Drawn from dried specimens in the Harbarium of the Royal Botanical
Gardens at Kew.

a filbert, but shining, and mottled with brown upon grey in the manner of castor-oil seeds. They are agreeable to the taste, and are stored up by the Indians, who experience no inconvenience from eating them, notwithstanding their relation to such acrid plants as the West Indian purging nut, or the Tiglium bushes of the East Indies. Aublet himself was able to eat them in abundance without inconvenience.

" It was long supposed that there was only one species of Siphonia ; but the enterprise of Mr. Spruce, a distinguished naturalist, now engaged in investigating Brazil, has made botanists acquainted with several others, viz. : —

Siphonia lutea, Bentham—found near Panuré, on the river Uaupes.

Siphonia discolor, Spruce—from the north bank of the Amazon at its junction with the Rio Negro, and also from Panuré.

Siphonia paucifolia, Spruce—from Panuré.

Siphonia rigidifolia, Spruce—from Panuré.

Siphonia Spruceana, Bentham—from Santarem in the province of Para.

" II. *Hancornia Speciosa.* — Under the name of Mangaba or Mangava the Brazilians have a fruit the produce of an Apocynaceous plant, to which botanists have given this appellation. It is very common about Pernambuco and Olinda, and also occurs at Bahia. Gardner describes it as ' reaching to the size of an ordinary apple tree, though its small leaves and drooping branches give it more the appearance of the weeping birch. The fruit is yellow, a little streaked with red on one side, about the size of an Orleans plum, and of delicious flavour. When in season, it is brought in great quantities to Pernambuco for sale.' According to Mr. Claussen, the tree is found 'on the high plateaux of South America, between 10° and 12° S. lat., at a height from 3000 to 5000 feet above the sea. The leaves are opposite each other on the slender branches, about two inches long, oblong, suddenly ending in a blunt point, shining above, but on the under side pale, with fine parallel veins. The flowers are slender, tubular, about one-and-a-half inches long, and grow singly from among the leaves.' In his communication to the British Association in 1855, Mr. Claussen stated that the Hancornia belonged to the same Sapotaceous order as the gutta percha tree, but in this he was mistaken. We have no further particulars concerning this tree, which yields Pernambuco caoutchouc.

" III. *Ficus elastica.*—This familiar Indian plant, now so common in green-houses, is, according to Roxburgh, known by the name of Kasmeer to the inhabitants of the Pundua and Juntipoor mountains, which bound the province of Silhet on the north; it is there found wild, and forms a tree fully as large as the sycamore in England. The trunk is said to be from five to six or more feet in circumference, the wood soft, porous, of a light brown colour, and only fit for fuel or charcoal. Its branches are numerous, spreading and rising

Thomas Hancock Jun.t, Delt.   Alfred S. Petter, Lith.

FICUS ELASTICA.

Assam.

Tho.ͤ Hancock Jun.ͬ Delt

Alfred.S.Petter, Lith.

URCEOLA ELASTICA.

Copied from a Print (by Roxburgh) in the Asiatic Researches.

in every direction, forming a very extensive and shady head.   The figs grow in pairs from the base of the broad, leathery, shining, deep green leaves; when ripe they are oval, about the size of an olive, smooth, and of a greenish-yellow colour.   They do not appear to be eaten.

" Roxburgh says, that with the milk, while in its recent state, ' the natives of the mountains, a most barbarous race as can be found in any part of the world, pay the inside of their rude utensils, that are intended to hold fluids ; the caoutchouc itself, being very inflammable, furnishes them with candles and flambeaux.'   Roxburgh found it perfectly soluble in Cajuputi oil.

" IV. *Urceola elastica.*— Mr. James Howison, a surgeon residing in Prince of Wales's Island, is the authority for the following fact, taken from a memoir published by him in the year 1798, in the fifth volume of the ' Asiatic Researches.'   While clearing a way through jungle with cutlasses, it was remarked that a vine had been divided, the milk of which, drying on the blade of the weapon, possessed all the properties of American caoutchouc.   The vine was about as thick as a man's arm, with a strong cracked ash-coloured bark.   It had joints at a small distance from each other, often sent out roots, seldom branches, ran along the ground to a great length, and at last rose upon the highest trees into the open air.   It was found in the greatest plenty at the foot of the mountains, upon a red clay mixed with sand, in situations completely shaded. It was afterwards met with on the west coast of Sumatra, and other Malay countries.

" Roxburgh describes its leaves as being opposite, on short stalks, oblong, pointed, a little rough, with a few scattered white hairs on the underside. The flowers are small, of a dull greenish colour, and are produced at the ends of the shoots in bunches, like those of a lilac bush. The seed-vessel is laterally compressed into the form of a turnip, is wrinkled, leathery, about three inches in the greatest diameter. The seeds are very numerous, and immersed in pulp.

" Like Hancornia, this plant belongs to the Apocynaceous order, or race of Dogbanes."

As the following letter (which we have received from our correspondents in Para) contains some additional and interesting information on the proceedings of the collectors of the liquid rubber, and their modes of manufacturing it into bottles, and other forms, I here insert it: —

" Per Steamer. Para, 1st July, 1856.
" Gentlemen,
. . . . . " We can give you to-day the following brief outline of the way in which india-rubber is manufactured.

" The Indians unite together generally in a pretty good number, and proceed to discover some

spot in the virgin forest where there are rubber
trees. As soon as they have found such a place,
they cut paths through the wood to it. This is
the sole difficulty experienced in procuring rubber,
but it is a great one, as, owing to the fertility of
the soil, the vegetation forms an almost closed
mass, and every step must be gained by the axe.
As soon as this labour is accomplished, they make
an incision in the tree, at the height of a man's
body from the ground, and arrange rude bowls
of clay which hold about a tumbler full, stick
the bowls to the trees a little below the incision,
and collect therein the milk running out; such
a bowl is filled in about three hours, if the tree
be fruitful. When the first cutting ceases to
yield, they make a second one the same distance
lower down, and so on until they have exhausted
the milk in the tree, which is done by making in
all four incisions, all at equal distances ; they then
pour the milk into larger vessels, gather heaps
of Urucari or Inaja nuts, which yield a thick oily
smoke, and set them on fire; they now begin the
manufacturing process by covering the wooden
forms for sheets, long and flat bottles, &c. with
clay (in order to be able to detach the rubber
easily afterwards), dip the forms into the milk,
and hold them over the smoke. As soon as the
milk is dry, they dip them a second time, and so
on till the rubber is of sufficient thickness ; they
then take it off the form, and the rubber is ready
for exportation. All rubber is manufactured in

156 ORIGIN OF THE INDIA-RUBBER MANUFACTURE.

this manner, the difference in quality depending upon the greater or lesser amount of clay and dirt which has become mixed with the milk. The first manufactured is the best (fine); and the last, made of milk adulterated with clay which has fallen from the different forms already dipped in, is the worst. A tree cannot be again made use of for two years, as it requires that time to recover its exhausted strength. There is another way of getting the milk, which is, however, forbidden by Government, as it destroys the tree. This is, to bind the tree at the top and bottom with willow twigs, and then draw off all the milk at once with incision.

" In the smoking process they have tried different qualities of coals and woods, but without use. Perhaps you observed already small lots of not smoked fine rubber which they sell here as mixed; this rubber comes from the interior of the Amazon province, where they don't have the above-mentioned fruits, and in consequence cannot smoke the rubber perfectly. All Indians give the preference to the nuts.

" We are, Gentlemen,

" Your obedient servants, &c. &c."

TABLES

OF

EXPORTS AND IMPORTS.

———————

THE annexed Tables, exhibiting the exports of india rubber from Para from 1836 to 1855 inclusive ; the imports to, and the exports from, Singapore from 1849 to 1855 ; and the imports to, and the exports from, the United Kingdom from 1842 to 1855, are all specially compiled for this work from official documents, involving considerable expense, and it is hoped that they will not be copied without acknowledgment.

Ports of Destination.	1836 to 1837.			1837 to 1838.			1838 to 1839.		
	Native Shoes.	Different Shapes.		Native Shoes.	Different Shapes.		Native Shoes.	Different Shapes.	
		Fine and mixed.	Coarse and Sernamby.		Fine and mixed.	Coarse and Sernamby.		Fine and Mixed.	Coarse and Sernamby.
	Pairs.	lbs.	lbs.	Pairs.	lbs.	lbs.	Pairs.	lbs.	lbs.
Antwerp	...	...	...	800	3,714	...	480	19,448	...
Baltic	...	...	...	...	...	...	...	...	...
Baltimore	...	...	...	...	...	...	...	...	...
Barbadoes	...	...	•	...	...	...	...	...	...
Barcelona	...	...	...	...	...	...	...	,,.	...
Bordeaux	...	...	...	...	...	...	...	...	...
Boston	7,654	5,165	...	23,813	7,003	...	17,694	2,176	...
Bremen	...	...	...	...	...	...	...	...	...
Cayenne	514	...	...	...	3,286	...	850	6,268	...
Copenhagen	...	...	...	...	...	...	...	...	...
Cowes	...	...	...	...	...	...	...	...	...
Falmouth	...	...	...	...	...	...	...	...	...
Genoa	...	...	...	...	...	...	...	6,206	...
Gibraltar	...	...	...	...	672	...	117	...	...
Guadaloupe	...	...	...	...	...	...	...	5,446	...
Hamburg	3,605	3,104	...	22,349	11,228	...	11,691	18,216	...
Hàvre	...	...	...	...	16,553	...	...	...	...
Leghorn	12,860	125	...	...	...	...	...	...	...
Lisbon	...	43,489	...	...	44,555	...	2,440	26,406	...
Liverpool	...	...	...	...	...	...	...	...	...
London	363	25,321	...	11,717	71,271	...	16,631	111,087	...
Marseilles	600	24,336	...	2,659	10,501	...	...	...	...
Nantes	...	...	...	1,304	199	...	...	...	...
Newhaven	...	...	...	...	...	...	...	...	...
New York	35,561	12,387	...	52,335	3,756	...	...	...	...
Oporto	...	...	...	...	96	...	...	...	...
Salem	69,822	27,808	...	97,486	20,753	...	77,982	8,946	...
Surinam	..	...	...	...	...	...	300	...	...
Trieste	...	...	...	...	...	...	...	...	...
	130,979	141,735	...	212,463	193,587	...	128,185	204,199	...

Ports of Destination.	1839 to 1840.			1840 to 1841.			1841 to 1842.		
	Native Shoes.	Fine and mixed.	Coarse and Sernamby.	Native Shoes.	Fine and mixed.	Coarse and Sernamby.	Native Shoes.	Fine and Mixed.	Coarse and Sernamby.
	Pairs.	lbs.	lbs.	Pairs.	lbs.	lbs.	Pairs.	lbs.	lbs.
Antwerp	...	...	...	282	5,536	...	...	...	...
Baltic	...	...	...	...	...	...	...	...	...
Baltimore	...	...	...	...	...	...	...	...	...
Barbadoes	...	...	...	...	...	...	...	...	...
Barcelona	19	...	...	...	...	...	...	6,979	...
Bordeaux	...	...	...	...	...	...	...	...	...
Boston	80,026	19,401	...	70,198	7,111	...	100,035	11,401	...
Bremen	...	...	...	...	...	...	...	...	...
Cayenne	...	...	...	1,146	5,808	...	...	...	...
Copenhagen	...	...	...	...	...	...	...	...	...
Cowes	...	...	...	...	...	...	...	...	...
Falmouth	...	...	...	...	...	...	...	...	...
Genoa	...	2,432	...	...	1,760	...	...	...	...
Gibraltar	580	17,056	...	...	...	...	...	...	...
Guadaloupe	...	...	...	...	...	...	...	...	...
Hamburg	9,889	26,038	...	2,690	28,736	...	9,500	23,764	...
Hâvre	16,021	133,016	...	1,662	20,592	...	2,598	31,976	...
Leghorn	...	...	...	...	...	...	...	...	...
Lisbon	...	72,346	...	1,500	138,098	...	520	25,194	...
Liverpool	...	...	...	564	22,869	...	...	29,365	...
London	732	241,354	...	587	178,891	...	...	40,996	...
Marseilles	...	18,479	...	...	296	...	1,340	6,368	...
Nantes	3,583	79,804	...	...	...	...	...	7,112	...
Newhaven	...	...	...	...	...	...	...	...	...
New York	27,403	2,568	...	96,878	50,656	...	123,080	37,965	...
Oporto	...	358	...	439	960	...	2,272	5,034	...
Salem	96,127	18,048	...	141,341	49,698	...	239,115	19,114	...
Surinam	100	...	...	...	...	...	...	...	...
Trieste	...	...	...	...	...	...	...	...	...
	234,483	630,900	...	317,287	511,011	...	478,460	245,265	...

Ports or Destination.	1842 to 1843.			1843 to 1844.			1844 to 1845.		
	Native Shoes.	Different Shapes.		Native Shoes.	Different Shapes.		Native Shoes.	Different Shapes.	
		Fine and Mixed.	Coarse and Sernamby.		Fine and Mixed.	Coarse and Sernamby.		Fine and Mixed.	Coarse and Sernamby.
	Pairs.	lbs.	lbs.	Pairs.	lbs.	lbs.	Pairs.	lbs.	lbs.
Antwerp	692	10,816	...	350	7,552	...	...	...	...
Baltic	...	...	...	...	...	...	...	...	...
Baltimore	...	...	...	...	...	...	...	...	...
Barbadoes	...	...	...	...	...	...	...	...	...
Barcelona	...	...	...	...	...	...	...	11,360	...
Bordeaux	...	...	...	6,410	4,182	...	...	...	...
Boston	99,997	13,156	...	78,042	37,501	...	57,185	41,920	...
Bremen	...	...	...	...	...	...	...	...	...
Cayenne	1287	8,400	...	...	...	...	...	...	...
Copenhagen	...	...	...	...	...	...	...	...	...
Cowes	18,700	5,856	...	...	...	...	...	...	...
Falmouth	...	...	...	...	...	...	...	...	...
Genoa	...	...	...	102	18,144	...	...	...	...
Gibraltar	1,000	...	...	4,667	1,568	...	1,385	...	...
Guadaloupe	...	...	...	...	...	...	...	...	...
Hamburg	8,762	8,848	...	70,033	27,496	...	11,307	13,568	...
Hâvre	10,182	21,456	...	7,334	10,240	...	17,290	22,516	...
Leghorn	...	...	...	...	...	...	...	...	...
Lisbon	15,925	9,592	...	962	2,518	...	...	3,520	...
Liverpool	...	28,172	...	...	...	...	...	18,192	...
London	26,492	...	...	4,902	105,202	...	...	131,696	...
Marseilles	2,600	13,024	...	4,298	3,200	...	200	12,960	...
Nantes	512	9,626	...	17,071	17,008	...	8,586	21,472	...
Newhaven	...	...	...	...	...	...	...	...	...
New York	80,141	13,156	...	108,223	39,488	...	129,403	140,176	...
Oporto	1,193	1,376	...	800	256	...	...	488	...
Salem	119,339	15,466	...	99,871	46,400	...	189,982	129,408	...
Surinam	...	...	...	...	...	...	...	...	...
Trieste	...	...	...	...	...	...	...	...	...
	386,822	158,944	...	403,065	320,755	...	415,338	547,276	...

Ports of Destination.	1845 to 1846.			1846 to 1847.			1847 to 1848.		
	Native Shoes.	Different Shapes.		Native Shoes.	Different Shapes.		Native Shoes.	Different Shapes.	
		Fine and Mixed.	Coarse and Sernamby.		Fine and Mixed.	Coarse and Sernamby.		Fine and Mixed.	Coarse and Sernamby.
	Pairs.	lbs.	lbs.	Pairs.	lbs.	lbs.	Pairs.	lbs.	lbs.
Antwerp	625	...	...	1,341	7,872	...	...	...	...
Baltic	14,555	10,720	...	...	...	...	...	...	...
Baltimore	...	3,200	...	...	...	...	...	...	...
Barbadoes	...	...	...	...	...	...	...	...	...
Barcelona	...	...	...	...	...	...	...	...	...
Bordeaux	2,220	10,208	...	...	...	...	...	...	...
Boston	3,386	21,168	...	31,654	8,992	...	...	...	...
Bremen	...	...	...	18,193	2,048	...	...	...	...
Cayenne	200	...	...	...	...	...	476	...	...
Copenhagen	...	...	...	...	...	...	300	1,152	...
Cowes	2,740	4,256	...	...	...	...	...	...	...
Falmouth	...	20,928	...	...	...	...	...	...	...
Genoa	...	11,808	...	55	3,328	...	...	5,376	...
Gibraltar	...	...	...	...	...	...	...	...	...
Guadaloupe	...	...	...	...	...	...	...	...	...
Hamburgh	51,757	62,752	...	11,332	5,824	...	9,383	69,616	...
Hâvre	1,614	21,664	...	13,419	25,720	...	400	13,312	...
Leghorn	...	...	...	...	...	...	...	...	...
Lisbon	883	3,248	...	2,030	10,912	...	2,635	2,160	...
Liverpool	...	32,320	...	9,062	181,248	...	...	176,864	...
London	8,354	68,864	...	...	112,160	...	...	45,824	...
Marseilles	6,500	33,728	...	4,234	992	...	...	6,592	...
Nantes	25,700	68,096	...	24,078	4,064	...	4,600	40,881	...
Newhaven	...	...	...	...	...	...	...	...	...
New York	168,111	222,208	...	207,219	406,112	...	110,964	443,424	1,984
Oporto	2,445	...	...	1,044	...	...	200	...	...
Salem	127,008	186,238	...	107,128	237,152	...	108,421	216,416	...
Surinam	...	...	...	...	...	...	...	...	...
Trieste	...	...	...	100	...	...	...	...	...
	416,078	781,406	...	430,889	1,006,424	...	237,379	1,021,617	1,984

M

Ports of Destination.	1848 to 1849.			1849 to 1850.			1850 to 1851.		
	Native Shoes.	Different Shapes.		Native Shoes.	Different Shapes.		Native Shoes.	Different Shapes.	
		Fine and Mixed.	Coarse and Sernamby.		Fine and Mixed.	Coarse and Sernamby.		Fine and Mixed.	Coarse and Sernamby.
	Pairs.	lbs.	lbs.	Pairs.	lbs.	lbs.	Pairs.	lbs.	lbs.
Antwerp	...	704	...	265	...	...	233	6,496	...
Baltic	...	...	...	...	...	...	...	...	...
Baltimore	...	...	...	...	...	...	...	...	...
Barbadoes	...	...	...	...	32	...	...	...	...
Barcelona	...	...	...	...	1,856	...	...	...	...
Bordeaux	...	...	...	...	...	...	...	...	...
Boston	...	...	...	6,217	50,912	...	...	...	...
Bremen	...	...	...	...	...	...	...	...	...
Cayenne	...	...	...	...	...	...	...	...	...
Copenhagen	...	...	...	...	...	...	...	...	...
Cowes	...	...	...	...	...	...	...	...	...
Falmouth	1,200	10,784	...	...	...	...	...	...	...
Genoa	...	...	...	...	15,136	...	...	...	...
Gibraltar	...	...	...	...	...	...	...	...	...
Guadaloupe	...	...	...	...	...	...	...	...	...
Hamburg	...	...	...	12,155	24,544	...	31,384	90,656	...
Hâvre	...	87,968	...	600	74,944	...	6,470	69,664	...
Leghorn	...	...	...	...	...	...	...	..	...
Lisbon	...	6,112	...	35,286	512	...	7,695	1,264	...
Liverpool	...	384,544	2,816	8,532	244,400	...	12,350	899,104	17,948
London	...	64,560	...	...	18,176	...	8,495	188,000	17,920
Marseilles	210	6,688	...	...	28,256	...	400	4,512	...
Nantes	5,725	134,528	...	6,074	87,776	...	120	154,880	...
Newhaven	...	...	...	...	...	...	...	...	...
New York	190,535	730,496	2,656	165,254	656,480	...	16,550	743,135	6,048
Oporto	57	288	...	...	96	...	7,618	...	...
Salem	115,898	176,640	...	80,496	530,816	...	47,557	738,464	6,720
Surinam	...	...	...	...	...	...	...	...	...
Trieste	...	...	...	...	...	...	...	...	...
	313,625	1,603,312	5,472	314,879	1,733,936	...	138,872	2,896,175	48,672

Ports of Destination.	1851 to 1852.			1852 to 1853.			1853 to 1854.		
	Native Shoes.	Different Shapes.		Native Shoes.	Different Shapes.		Native Shoes.	Different Shapes.	
		Fine and Mixed.	Coarse and Sernamby.		Fine and Mixed.	Coarse and Sernamby.		Fine and Mixed.	Coarse and Sernamby.
	Pairs.	lbs.	lbs.	Pairs.	lbs.	lbs.	Pairs.	lbs.	lbs.
Antwerp	511	11,200	...	2,567	...	...	...	...	...
Baltic	...	...	...	...	...	...	...	...	...
Baltimore	...	...	...	...	•••	...	...	...	...
Barbadoes	...	...	...	...	...	...	...	...	...
Barcelona	...	...	...	...	...	...	...	...	...
Bordeaux	...	...	...	...	...	...	...	...	...
Boston	...	...	...	...	...	...	...	83,888	52,302
Bremen	...	...	...	...	...	...	...	...	...
Cayenne	...	...	...	...	...	...	...	...	...
Copenhagen	...	...	...	...	...	•••	...	...	...
Cowes	...	...	...	...	...	...	...	...	...
Falmouth	...	...	...	...	...	...	...	...	...
Genoa	...	1,408	...	...	...	...	...	...	...
Gibraltar	...	5,120	...	...	...	...	...	...	...
Guadaloupe	...	...	...	...	...	...	...	...	...
Hamburg	21,948	124,352	...	15,187	38,496	5,976	...	32,106	17,378
Hâvre	...	192,800	...	...	57,452	7,008	...	66,704	24,790
Leghorn	...	...	...	...	...	...	...	...	...
Lisbon	42,934	14,037	...	7,145	2,140	...	22	262	...
Liverpool	...	906,449	208,384	8,354	299,478	828,520	...	578,530	792,445
London	...	291,936	38,400	...	54,624	19,334	...	40,103	56,847
Marseilles	600	20,704	...	...	...	...	...	19,824	2,126
Nantes	...	92,256	...	...	12,586	...	...	20,992	25,880
Newhaven	...	143,808	1,952	...	110,048	50,016	...	144,185	63,220
New York	...	899,889	14,520	6,875	736,992	572,208	...	1,636,852	638,212
Oporto	1,503	2,184	...	300	64	...	...	68	...
Salem	19,283	389,643	16,320	39,455	520,316	166,242	15,787	517,504	232,784
Surinam	...	...	...	...	...	...	...	...	...
Trieste	...	...	...	...	...	•••	...	...	...
	86,779	3,095,786	279,576	79,883	1,832,196	1,649,304	15,809	3,141,018	1,905,984

Ports of Destination.	1854 to 1855.			1855 to 1856.		
	Native Shoes.	Different Shapes.		Native Shoes.	Different Shapes.	
		Fine and Mixed.	Coarse and Sernamby.		Fine and Mixed.	Coarse and Sernamby.
	Pairs.	lbs.	lbs.	Pairs.	lbs.	lbs.
Antwerp	...	16,518	...	...	5,378	...
Baltic	...	...	...	...	...	...
Baltimore	...	...	...	...	...	...
Barbadoes	...	...	...	...	...	...
Barcelona	...	...	...	...	...	...
Bordeaux	...	...	...	...	...	...
Boston	...	287,010	133,582	...	228,668	5,557
Bremen	...	...	...	...	...	...
Cayenne	...	...	...	...	...	...
Copenhagen	...	...	...	...	...	...
Cowes	...	...	...	...	...	...
Falmouth	...	...	...	...	...	...
Genoa	...	...	...	...	...	...
Gibraltar	...	...	...	...	...	...
Guadaloupe	...	...	...	...	...	...
Hamburg	...	114,780	21,145	...	56,884	6,952
Hâvre	...	112,005	2,655	...	59,075	5,214
Leghorn	...	...	...	...	...	...
Lisbon	...	4,883	...	...	5,729	6,120
Liverpool	...	1,079,401	976,779	...	969,919	739,877
London	...	62,866	29,436	...	80,492	141,722
Marseilles	...	...	...	...	32	...
Nantes	...	23,694	5,594	...	43,679	790
Newhaven	...	231,820	52,016	...	107,897	7,104
New York	...	1,448,103	535,834	...	1,417,332	45,635
Oporto	...	...	...	...	280	288
Salem	...	387,976	196,802	...	502,080	17,074
Surinam	...	...	...	...	...	...
Trieste	...	...	...	...	...	...
	...	3,769,056	1,953,843	...	3,477,445	976,333

Ports of Destination.	Total, 1836 to 1856.		
	Native Shoes.	Different Shapes.	
		Fine and Mixed.	Coarse and Sernamby.
	Pairs.	lbs.	lbs.
Antwerp	8,146	95,234	...
Baltic	14,555	10,720	...
Baltimore	...	3,200	...
Barbadoes	...	32	...
Barcelona	19	20,192	...
Bordeaux	8,610	14,390	...
Boston	575,904	825,472	191,441
Bremen	18,193	2,048	...
Cayenne	4,473	23,762	...
Copenhagen	300	1,152	...
Cowes	21,440	10,112	...
Falmouth	1,200	31,712	...
Genoa	157	65,598	...
Gibraltar	7,749	24,416	..
Guadaloupe	...	5,446	...
Hamburg	302,972	781,008	51,451
Hâvre	77,490	1,037,657	39,667
Leghorn	12,860	125	...
Lisbon	119,977	416,977	6,120
Liverpool	38,862	5,822,683	3,566,805
London	78,273	1,961,579	303,659
Marseilles	23,741	210,492	2,126
Nantes	97,353	818,653	32,264
Newhaven	...	737,758	174,308
New York	1,518,532	9,641,175	1,817,097
Oporto	17,871	11,548	288
Salem	1,792,098	4,738,886	635,942
Surinam	400	...	...
Trieste	100	...	...
	4,741,275	27,312,027	6,821,168

Imports of India-Rubber to the United Kingdom for the Years 1842 to 1855 inclusive.

IMPORTS.

From	1842.	1843.	1844.	1845.	1846.	1847.	1848.	1849.
	lbs.	lbs.	lbs.	lbs.	lbs.	lbs.	lbs.	lbs.
East Indies -	42,112	7,504	...	...	1,568	13,776	45,472	62,608
United States -	33,936	17,024	13,664	1,232	55,552	2,576	...	...
Brazil - -	222,432	306,320	422,576	329,952	440,272	630,336	417,200	515,760
Java - -	...	27,664	2,240	...	224	11,760	6,608	8,400
Elsewhere -	18,704	784	10,416	9,408	54,096	1,120	2,016	9,968
Total	317,184	359,296	448,896	340,592	551,712	659,568	471,296	596,736

From	1850.	1851.	1852.	1853.	1854.	1855.	Total.
	lbs.	lbs.	lbs.	lbs.	lbs.	lbs.	lbs.
East Indies -	32,480	66,864	356,272	391,216	663,936	940,128	2,623,936
United States -	61,488	181,888	...	21,392	277,200	284,928	950,880
Brazil - -	668,304	1,237,936	1,435,056	1,143,520	1,660,960	2,395,344	11,825,968
Java - -	35,504	191,968	293,888	77,230	184,912	203,304	1,060,752
Elsewhere -	55,328	31,472	110,768	307,104	302,848	1,166,032	2,080,064
Total	858,104	1,710,128	2,195,984	1,940,512	3,089,856	5,006,736	18,541,600

Exports of India-Rubber from the United Kingdom for the Years 1842 to 1855 inclusive.

EXPORTS.

Destination.	1842.	1843.	1844.	1845.	1846.	1847	1848.	1849.
	lbs.	lbs.	lbs.	lbs.	lbs.	lbs.	lbs.	lbs.
Russia ...	8,624	35,392	10,080	15,904	18,704	54,320	77,616	87,472
Hanse Towns	13,440	11,536	26,880	13,216	3,696	24,528	224	32,368
Holland ...	6,272	43,232	29,792	37,184	112	26,432	112	61,040
Belgium ...	4,592	38,752	74,816	120,736	13,552	4,032	...	3,024
France ...	...	1,008	22,064	448	...	224	...	...
United States	112	...	...	...	...	18,256	...	...
Elsewhere ...	672	2,576	1,232	...	...	6,384	896	2,688
Total	33,712	132,496	164,864	187,488	36,064	134,176	78,848	186,592

Destination.	1850.	1851.	1852.	1853.	1854.	1855.	Total.
	lbs.	lbs.	lbs.	lbs.	lbs.	lbs.	lbs.
Russia ...	39,088	...	28,784	14,896	...	...	390,880
Hanse Towns	49,392	57,904	52,528	25,038	163,520	300,496	774,816
Holland ...	6,160	60,704	88,592	122,640	72,464	5,936	560,672
Belgium ...	22,176	57,344	232,512	290,752	135,856	36,400	1,034,544
France ...	560	...	...	...	193,984	475,776	694,064
United States	...	168,448	189,616	210,784	638,736	103,040	1,328,992
Elsewhere	...	2,576	5,488	6,384	13,104	13,552	55,552
Total	117,376	346,976	597,520	670,544	1,217,664	935,200	4,839,520

M 4

Imports and Exports of India-Rubber at Sincapore, for the Years
1849–50 to 1854–55.

IMPORTS.

From	1849 1850	1850 1851	1851 1852	1852 1853	1853 1854	1854 1855	Total.
	lbs.	lbs.	lbs.	lbs.	lbs.	lbs.	lbs.
Java	20,608	17,248	37,856	175,616	57,904	49,504	358,736
Sumatra	1,344	448	...	194,096	139,104	428,288	763,280
China	...	1,680	...	...	...	...	1,680
Manilla	...	14,896	...	...	...	...	14,896
Borneo	...	...	...	2,688	224	112	3,024
Malay Peninsula ...	...	...″	...	448	...	...	448
Penang and Malacca	...	...	...	...	127,120	126,448	253,568
Elsewhere	...	...	...	96,544	...	33,712	130,256
Total lbs.	21,952	34,272	37,856	469,392	324,352	638,064	1,525,888

EXPORTS.

Destination.	1849 1850	1850 1851	1851 1852	1852 1853	1853 1854	1854 1855.	Total.
	lbs.	lbs.	lbs.	lbs.	lbs.	lbs.	lbs.
Great Britain ...	21,840	37,856	109,200	538,832	449,120	390,768	1,547,616
North America ...	21,280	6,832	3,136	18,592	133,952	341,040	524,832
France	...	...	...	10,080	63,392	42,560	116,032
Hamburg	...	...	...	...	...	60,032	60,032
Java	...	...	...	...	...	1,344	1,344
Total lbs.	43,120	44,688	112,336	567,504	646,464	835,744	2,249,856

The above Statement shows an Export during six years, in excess of the Imports, of
723,968 lbs., the produce of Sincapore.

MECHANICAL APPLICATIONS OF
VULCANIZED INDIA-RUBBER.

Note.—A, Quality is the most elastic. It weighs about 60 lbs. per cubic foot, or 1-29th of a lb. per cubic inch.

D, Quality weighs about 82 lbs. per cubic foot, or 1-21th of a lb. per cubic inch.

E, Quality more elastic than D—weighs about 92 lbs. per cubic foot, or 1-19th of a lb. per cubic inch.

F, C.—Fibrous compound, used for flange washers, valves, and pump buckets. Weight 1-25th of a lb. per cubic inch.

THE following list describes some of the applications of india-rubber. Many of these articles are formed of *pure* vulcanized rubber, and others prepared with various pigments according to the required colour, quality, or intended application of the article, each modification of quality being distinguished by a letter, thus enabling the consumer to select either *pure* vulcanized rubber or any of the stated compounds.

BUFFER AND BEARING SPRINGS. (*Fuller and D'Bergue's Patent.*)

A patent application of the vulcanized india-rubber to the purposes of draw springs, buffers, and bearing springs of railway carriages; more efficient, durable, and economical than any modification of steel springs for such purposes.

CYLINDERS.

These are made of any dimensions of bore and thickness; they are supplied for the formation of buffers, washers, and springs, where enormous compression is used, and to relieve the concussion of steam hammers, fulling mills, &c. &c.

FOOT AND PUMP VALVES FOR OCEAN STEAMERS.

These valves are rapidly superseding the metallic
valves; concussion is avoided; a perfect joint formed
under the most ·rapid motion of a steam engine;
they are extensively used in steam ships constructed
with paddle wheels; for screw steamers they are
quite indispensable.

VALVE CANVASS.

This is a modification of the above; it is prepared for
various mechanical purposes, and is much less elastic
than the vulcanized india-rubber, in which canvass
or other fibrous materials are not incorporated.

ENGINE PACKING.

This article, being a compound of india-rubber and
fibrous materials, is eminently adapted for packing
pistons, stuffing boxes, and the various parts of steam
engines that require packing; it is supplied in sheets,
slips or rings, and the fibres are so arranged in the
compound as to give the greatest possible amount of
durability and relief from friction.

WASHERS FOR FLANGE AND SOCKET JOINTS.

By means of these washers an instantaneous joint
may be made for every purpose, and in every con-
ceivable situation in manufacturing establishments;
they are made to any figure or size, from that of
the smallest pipe to the largest chemical cistern.

WHEEL TIRES.

For this purpose the vulcanized india-rubber is firmly
attached to a metal hoop or tire, of the usual width,
or an endless band of rubber is sprung on to an
ordinary wheel tire, and kept in position by a flange
on either side; the rubber, projecting from the
flanges, rests on the ground, and this prevents the
concussion to which the wheels are ordinarily subjected, and altogether increases the durability of the

carriage. Carriages having these tires roll along without the slightest noise, and in an extraordinarily soft and easy manner. They are particularly valuable for trucking goods in warehouses, railway stations, and for bath and invalid chairs.

ROLLING PISTON FOR LIFTING AND FORCING PUMPS. (*Woodcock's Patent.*)

A patent application of the vulcanized india-rubber. The rolling pistons prevent all friction in pump barrels and water meters, and they are economical, enduring, and cannot "choke," whatever be the fluid they are employed to pump.

RINGS, STRIPS, AND CORDS FOR ELASTIC PURPOSES.

These may be applied to the most delicate or most powerful mechanical purposes.

FLEXIBLE PUMPS FOR FORCING OR EXHAUSTING AIR, GASES, &c. &c.

PUMP BUCKETS.

These are made of various forms, and their advantages are, durability, and the facility with which water, liquid manures, and chemical liquids of any temperature may be pumped without injury to the vulcanized india-rubber.

PLUG VALVES.

These are conical valves for ships, chemical and water cisterns, and as plugs for hot or cold water baths; they never corrode, or fail in forming a perfect joint.

COACH LOOPS OR ROUND ROBBINS.

These are used by coachmakers instead of those of combined iron and leather, as being more safe, quiet, and durable for the support of carriage bodies.

TEAGLE OR HOIST STRAPS.

These are more enduring, safe, and economical than the ordinary hempen rope.

GAS BAGS.

A useful and certain apparatus for the repair or alteration of gas mains; by its aid a town need no longer be placed in darkness during the laying of pipes.

DIAPHRAMS.

These are made of vulcanized india-rubber, for dry gas meters, and for measuring the supply of water to towns, manufactories, or private dwellings.

DOOR SPRINGS.

These are used in a variety of ways: one of its simplest applications is, as a strong loop slipped over two hooks, one in the door and one in the jamb or door frame.

CORRUGATED RUBBER FELT.

This article is extensively used for manufacturing and railway purposes; in the latter, its peculiar properties are exhibited when placed under the chairs or flat rails; concussion is prevented, and wear and tear of the rails and carriages much diminished; it is also used extensively for the bottoms of vulcanized india-rubber over-shoes.

HOSE PIPES AND TUBING.

These are made of any bore and length, suited to the delivery hose and suction pipes of fire engines, and for the conveyance of gas, steam, acids, alkalies, and other fluids; they are also soft and pliable, and not injuriously affected by heat or cold.

LOCOMOTIVE PIPING.

This is a modification of the above class of articles; for this purpose the ordinary strength is increased.

ROLLERS FOR LETTER-PRESS PRINTING.

A substitute for the ordinary glue and treacle rollers. The chief recommendations are their permanent elasticity and durability.

BLANKETS FOR CALICO PRINTING.

These are formed of alternate layers of india-rubber and cloth; by their use, a printer can produce fine and delicate patterns, that could not result from the ordinary woollen blanket. These patent blankets are capable of printing 25,000 pieces of cloth, and the power required is much diminished, in consequence of their peculiar elasticity requiring less pressure to produce the pattern.

FURNISHERS FOR CALICO PRINTING.

These are formed of vulcanized india-rubber, with a roughened surface which takes up the colour, and applies it to the engraved roller. The same furnisher can readily be applied to any colour without waste; the composition of printing colours has no injurious effect on these furnishers, which perform their work for a series of years.

SIEVES OR FURNISHERS FOR SURFACE PRINTING.

These are found to be a great improvement on the ordinary woollen sieve, especially in an economical point of view, as they prevent the absorption of the colour.

ARTIFICIAL LEATHER FOR CARD BACKS.

This article is too well known to cotton and woollen manufacturers to need comment or explanation; suffice it to say, that its chief advantages are cheapness, evenness, large size of sheets, and elasticity.

MOULDED ARTICLES — ELASTIC.

Under this head is comprised the endless variety of forms in which the vulcanized india-rubber is produced for manufacturing, surgical, domestic, and fancy purposes, which may require moulds for their production. It is the peculiar property of vulcanized india-rubber to retain, permanently, the form in which it is vulcanized.

MOULDED ARTICLES — HARD VULCANIZED.

Under this head is included all the before-mentioned; the degree of hardness is adapted to the purposes to which the substance is required to be applied.

Hard vulcanized rubber is supplied in the form of sheets, slabs, bars, tubes, and can be moulded to any desired figure.

It is a substitute for bone, ivory, whalebone, hardwoods, &c., and is capable of being worked by the ordinary tools used for those substances.

It can be worked in the lathe, sawn, planed, drilled, screwed, or engraved.

The properties of this material are : — resistance to the action of hot and cold acids, alkalies or chemically impregnated solutions, the sulphur in gas; and it is suited to any uses in which metals are objectionable.

CUSHIONS FOR BILLIARD TABLES.

These are an improvement upon the original india-rubber cushions, which in cold weather were hard, and in that state useless.

The patent cushions never alter in their elasticity, and consequently are to be depended upon ; they are highly approved of by scientific players.

SEWER AND SINK VALVES.   (*Dr. C. Bell's Patent.*)

These effectually prevent the escape of effluvium in all situations ; they are simple in application, and the vulcanized india-rubber is not injured by contact with any fluids that may pass through the valve into the sewer.

CUMULATORS, OR ELASTIC POWER PURCHASES FOR PROJECTILE AND LIFTING APPARATUS, &c. (*Hodge's Patent.*)

Here the vulcanized india-rubber is used in the form of tubes or cords.   Simple or compound springs are brought to bear upon harpoons, arrows, balls, shot, and other missiles, and made to project them with immense velocity and precision. In case of lifting and suspending great weights, an assemblage of these springs are brought into use, by which a child may lift an enormous weight.

THREAD — VULCANIZED INDIA-RUBBER.

This article, by reason of its great strength and permanent elasticity, has greatly extended the trade in elastic woven and knitted fabrics. It is prepared of several degrees of fineness, and supersedes the original native, or common india-rubber thread.

VULCANIZED SHEET RUBBER.

These sheets are supplied in grey or black of any thickness from the 70th of an inch upwards, and varying from 5 to 50 yards in length, by 50 inches and upwards in width; from these may be cut bandages, springs, strips for joints, and linings for chemical and other vessels, &c.

RUBBER AND CLOTH COMBINED IN SHEETS.

This is commonly known as "insertion" rubber; where required, the cloth is made highly elastic, so as to stretch to the extent of the rubber; where this material is required not to stretch, a non-elastic cloth is used.

FINE CUT SHEET RUBBER.

These sheets are supplied to any thickness, and are capable of being joined up for the manufacture of various articles.

FINE SHEET RUBBER.

Made of any thickness from 36 inches wide and upwards, and from 5 to 50 yards long.

INDIA-RUBBER SOLUTION, OR VARNISH.

SURGICAL PURPOSES.

HYDROSTATIC BEDS.

These beds afford great relief to the afflicted —
facilitate their movement, and supply a soft support
to every part of the person.

IRON BEDSTEADS WITH ELASTIC SACKING.

These are adapted for hospitals and public insti-
tutions generally; they afford the greatest facility
for attendance on the sick; are much cheaper than
hydrostatic beds, and very simple in construction.

MATTRESSES, BEDS, AND PILLOWS.

(*With bellows for inflation.*)

These articles are prepared from air-proof materials,
and when inflated assume their figure of mattress,
bed, pillow, &c. ; they are very easy, light, and
portable, and made to any size and figure. Bellows
are supplied for the inflation of the larger articles,
and the operation is at once simple and easy.

ELASTIC WOVEN BANDAGES.

Are formed as stockings, knee-caps, leggings, thigh
pieces, anklets, armlets, wristlets, &c.; they are
extensively used for the relief of glandular swellings,
and varicose veins, abdominal belts, &c.; nothing
can be more easy and efficacious in use, or more
elegant in appearance, than these applications of the
vulcanized india-rubber thread.

BED SHEETS.

Applied as a cover, so that in cases of Hemorrhage,
&c., a valuable bed is entirely protected from injury.

WATER PILLOWS AND BEDS.

Are exceedingly elastic, either inflated for reclining
on, or for the application of hot or cold water to any
part of the body; to bed-ridden patients they are

invaluable, as they entirely prevent the friction produced by ordinary cushions or pillows.

CHEMICAL APRONS, SLEEVES, AND GLOVES.

Are used by surgeons in dissecting operations, and
prevent all risk from contact with poisonous fluids.
They are also valuable for manufacturing chemists,
dyers, and others, as they protect the person and
clothing from the action of caustic, alkalies, acids,
and other dangerous liquids.

GAS VESSELS.

Are made in the form of bags of any size or figure.
They are used for the purposes of illumination, and for
containing separate gases for chemical and experimental purposes; as, for instance, the oxy-hydrogen
microscope, &c.

TUBING, BRAIDED OR PLAIN.

For the conveyance of gas or other fluids, for
moveable lights and general use in manufactories,
chemical works, &c.

LIGATURES, &c.

These are made in the forms of thread, cords,
bandages, rings, &c., and are useful in cases of dis-
location, for the tourniquet and various other purposes
in surgical operations.

INJECTION BOTTLES, BREAST BOTTLES, EXHAUSTING
BELLS, ENEMAS, PESSARIES, URINALS, PIC NICS,
EAR PADS, TRUSS PADS, CORN PROTECTORS, FINGER
STALLS, &c.

Appliances for medical and surgical purposes: their
uses are indicated by their names, and are well
adapted to their varied purposes.

N

DOMESTIC APPLIANCES.

INFLATED CUSHIONS AND BEDS.

Are made of any form and to any dimensions: they are useful for chairs and sofas, or for the purposes of travelling—particularly in second class carriages on railways.

CHEST EXPANDERS.

Afford agreeable and healthful exercise to children and persons engaged in sedentary employment: they strengthen the muscular powers, expand the chest, and promote health. For schools and families they are particularly useful.

SPONGING BATHS.

A portable and efficient accompaniment of the modern bed-chamber: no water is absorbed by the material, and the bath can be packed away immediately after use; hence its value to persons travelling.

JAR COVERS AND CAPSULES.

For pickles, preserves, and anatomical specimens. They can be removed in a moment, and soundly replaced, afford protection from the atmosphere, and are perfectly self-fastening.

GUM RINGS, CORALS, NIPPLES.
CRIB SHEETS.
NURSING APRONS.
ARM GUSSETS.
SPONGE BAGS.
STRAPS FOR BABY JUMPERS.
DRESS DILATORS.
BATHING CAPS.
TOBACCO POUCHES.
BOTTLING CORKS AND BUNGS.
DECANTER STOPPERS.
TABLE MATS.

SEATS AND BACKS FOR CHAIRS AND STOOLS.
PLAYING BALLS.
FOOT BALLS.
CRICKET GLOVES AND BAT COVERS.
GLOVES OF ALL SIZES.

The names of the above articles indicate their use, and they are found safer and better adapted to their several uses than the articles they have superseded.

WEARING APPAREL.

PIECE GOODS, WATERPROOF FABRICS.
       CAMBRIC, SHEETING, LINEN, STUFF, ALPACA, SILK, WOOLLEN, &c. &c., DOUBLE AND SINGLE TEXTURE.
CAPES.
CAPES WITH SLEEVES OR LOOSE COATS.
CHESTERFIELD WRAPPERS.
COACHMEN'S COATS.
LADIES' PALETOTS.
BONNET HOODS WITH SHORT CAPE.
OVERALLS.
BRACES.
VEST BACKS.
GARTERS.
GAITERS.
TROUSER STRAPS.
OVER SHOES.
WEBBING FOR GUSSETS.
       "   BRACES.
       "   ELASTIC BOOTS.
       "   BANDAGES AND ROLLERS.
SANDLING.
APRON BANDS.
WRISTLETS.

LADIES' PAGES.
ELASTIC BELTS.
GLOVES.

Any explanation or statement of properties possessed by the above, is rendered almost superfluous by their general use by the public; but it may be observed, that of late years, the fashion has very much revived the use of waterproof clothing : the loose, open, easy character of the present dress renders a waterproof overcoat unobservable, made as they now are of materials such as are regularly worn in an unproof state, and consequently unobjectionable in appearance, whilst at the same time all the advantages of being kept perfectly dry in rain, is due to the india-rubber fabric alone.

NAUTICAL AND AGRICULTURAL ARTICLES.

SHIP SHEETS.

Are used for the purpose of passing under a ship's bottom in case of leakage or accident at sea. They effectually stop the ingress of the water, and enable the ship to proceed on her voyage without repairing the leak; on arriving at a port, the damage can be effectually repaired without taking the ship into dock, — advantages that have been appreciated by the Government and private traders.

SAFETY TUBES. (*Holdsworth's Patent.*)

Are used for life boats, life buoys, watching buoys, &c., and are peculiarly adapted for giving buoyancy to boats of all descriptions. They can be placed fore and aft, secured by nettings to the raisings or rails fitted for the purpose, or be secured across the boats under the thwarts, as the judgment of the owner may direct, thus converting any boat into a life boat. As life buoys, they may be thrown to the assistance of persons falling overboard: they are so light as not to do injury to the person who may be struck with them, while their buoyancy is such, that they are capable of supporting three persons in the water until assistance arrives.

LIFE BELTS, OR LIFE PRESERVERS.
Are made light, portable, and efficient for the protection of life and property in cases of wreck or other accidents at sea: many valuable lives have been saved by their use; and it is not too much to say, that in ordinary cases of shipwreck, the whole of the passengers and crew might be saved, if the properties of these valuable articles were more generally known.

BOATS INFLATED.
Are intended for pleasure purposes on lakes, for fishing, and for exploring parties going abroad. They can be packed in small compass while travelling, and, when required, can be converted into a boat in five minutes.

SOU'-WESTERS, DECK BOOTS, OVER SUITS, &c.
Are essential to officers, and sea-faring persons generally, as they preserve the usual clothing perfectly dry in the roughest weather.

DIVING DRESSES.
These entirely envelope the person, and enable divers and others engaged in submarine operations to perform their work in security.

CART, WAGGON, AND RICK COVERS.
Are thoroughly waterproof, very light, portable, and do not crack in use.

MALTING SHOES.

Are used by persons engaged in malting, and similar operations. They can be attached to common shoes, and enable persons to walk upon the grain without crushing or otherwise injuring it.

HOSE AND SUCTION PIPES.

For the conveyance of liquid manure, and the general purposes of land irrigation.

HORSE STOCKINGS, BRUSH AND RING BOOTS, KNEE CAPS, AND SHOE PADS.

These are found to answer the purpose much better than the ordinary leather ones, to which they are now preferred.

———

TRAVELLING AND SPORTING ARTICLES.

TRAVELLING CUSHIONS, BEDS, RUGS, BAGS, &c.
GIG APRONS.
DRIVING GLOVES.
CANTEENS AND BOTTLES.
DRINKING CUPS.
HORSE CLOTHS.
MUD BOOTS.
HANDLES FOR STICKS, UMBRELLAS, &c.

———

SHOOTING BOOTS.
GUN COVERS.
GAME BAGS.
SHOOTING HATS.
RIDING BELTS.
FISHING BOOTS.
FISHING STOCKINGS.
FISHING TROUSERS.
FISHING COATS.
YACHTING TROUSERS.
TENTS.

STATIONERY PURPOSES.

ELASTIC BANDS FOR PAPERS, LETTERS, &c.   (*Perry and Daft's Patent.*)
INKSTANDS WITH ELASTIC BOTTLE.
WRITING TABLETS.
PARCEL BANDS.
BOOK COVERS.
ERASING RUBBER IN SQUARES AND BOTTLES.

ORNAMENTAL.

ENAMELLED SHEETS.
EMBOSSED   "
MARBLED   "
PRINTS FROM ENGRAVED PLATES.
MAPS.
BAS RELIEFS.
MEDALLIONS.
EMBOSSINGS, FLOWERS, FIGURES, ANIMALS.
COLOURED THREAD, PLAIN AND BRAIDED.

Some of these articles exhibit the applicability of the vulcanized india-rubber to the Arts in cases where durable embossing of any degree of hardness, of fineness, of execution, susceptibility of colouring or elasticity are required.   Among the applications of embossing are the production of durable books for the BLIND, and both hard and elastic type for printing.

APPENDIX.

SPECIFICATIONS

OF

FOURTEEN PATENTS

GRANTED BY GEO. IV., WILL. IV., AND VICTORIA,
FROM APRIL 29. 1820, TO DEC. 30. 1847,

TO

THOMAS HANCOCK,
OF STOKE NEWINGTON,

FOR THE TREATMENT AND APPLICATION OF
INDIA-RUBBER.

O

SYNOPSIS.

With reference to Native Shoes, in the Statistical Tables, it should be mentioned that, in this country these shoes were mostly used as raw material in America, on the contrary, they were worn as over shoes.

APPENDIX.

SPECIFICATIONS.

ARTICLES OF DRESS.

Specification of Patent granted to THOMAS HANCOCK, *Stoke Newington, Middlesex, Esquire, for an Improvement in the Application of a certain Material to various Articles of Dress and other Articles, that the same may be rendered more Elastic.*—Dated April 29. 1820.

THE material I use is caoutchouc: I cut it into slips of a convenient length and thickness, according to the purpose for which it is to be used, and the degree of elasticity necessary. If the quality of the caoutchouc is not the best, or the spring is not required to be very substantial, I prepare those slips by putting them into hot water, and steeping them awhile, to prevent their cracking on the edges; when the substance of the spring is required to be more considerable, or the quality of the caoutchouc better, I use it without such preparation. I apply the caoutchouc spring to gloves in the following manner. A case or pipe of leather, linen, or cotton, or other similar material, is made, as long as it is necessary the spring should stretch; the spring is then fastened at the extremities of the pipe or case, by sewing or otherwise, in such a manner as that the pipe may contract or gather up very considerably. The case or pipe is then fixed in the wrist of the glove, so as to contract the glove to the size of the wrist, care being taken not to make the spring so strong but that the glove will easily draw over

P

the hand.   The case or pipe may be made in the glove
itself, and the spring introduced in the manner I have
described.   Attention must be paid in fastening the caoutchouc that it is not pierced anywhere between the extremities by the needle, otherwise it will be liable to tear and
break.   In a similar manner I apply the caoutchouc spring
to any article of dress where elasticity is desirable at any
particular part.   I apply the caoutchouc springs to waist-
coats and waistbands, to make them contract and sit close to
the body ; to coat-sleeve linings, to draw them closer round
the wrist; to the mouth of pockets, to prevent their contents from falling out when in an inverted position, and to
prevent their being easily picked; to trowser and gaiter
straps, to enable them to lengthen and shorten to the bend
of the knees and ankle-joints; to braces, instead of wire
and other springs, as now commonly used; to stockings, to
prevent their slipping down the leg ; to garters, to shirt-
wrists, to the knees of drawers and breeches ; to wigs, false
curls and fronts, to keep them tight on the head; to
pocket-books and purses, instead of the strap and loop, and
wire springs; to riding-belts, to stays, and such parts of
the apparel and dress of women as require to be kept close
to the person, and yet to be elastic as fastenings ; to boots,
shoes, clogs, and pattens, when the object is to take them off
and on without any lacing or tying.   I apply caoutchouc to
the soles of boots, shoes, and clogs, by making either the
whole sole of caoutchouc, or the inner or outer sole only, or
by fastening a piece of caoutchouc between the soles ; and
in either case boots, shoes, and clogs, are rendered more
elastic to the foot.   I apply the caoutchouc spring to stiff-
eners for neckcloths.   I use caoutchouc in stirrups to render
them elastic to the feet, by forming a piece to the bottom of
the stirrup, which I fasten on by having holes drilled in the
stirrup, and sewing on the caoutchouc with wax thread, or
wire, or by riveting or screwing it on with iron.   In this
specification I do not insist upon any particular mode of

applying or fastening the caoutchouc to the various articles described, as that may be varied as convenience may require; my object being to produce and apply a better kind of spring than any now in use for the purposes abovementioned. — In witness, &c.

THOMAS HANCOCK.

Enrolled August 8. 1820.

PITCH AND TAR.

Specification of Patent granted to THOMAS HANCOCK, *Stoke Newington, Middlesex, Esquire, for an Improvement in the Preparation for various useful purposes of Pitch and of Tar, separately or in union, by an Admixture of other Ingredients with either or both of them.* — Dated March 22. 1823.

MY invention consists in the mixing caoutchouc and its solvent with pitch or tar, or with pitch and tar combined together in various proportions, and thereby rendering these substances, namely, pitch and tar, whether separately or in combination with each other, less soluble in water, tougher, more elastic, and more durable than pitch or tar, or pitch and tar mixed together, and used in their natural state, are found to be. Previously to describing my method of preparing the solution of caoutchouc, &c., it is necessary for me to state that caoutchouc is the substance which is more generally known by the names elastic gum and Indian rubber. Many, if not all, of the essential oils will dissolve the Indian rubber; but for the sake of cheapness I prefer, and therefore employ, the essential oil of turpentine, or the essential oil of tar, either of which is a good solvent of the Indian rubber. I make the solution by first cutting the Indian rubber into very thin slips, in order to increase the quantity of surface of the said rubber and thereby to expedite its solution when it is immersed in the essential oil;

and I immerse it in the said essential oil either in a warm or
cold state, and occasionally stir it until the solution of the
Indian rubber is effected. I find by experience that the
solution is effected in a shorter time by exposing the essential oil and Indian rubber to a moderate degree of heat, that
is, not exceeding 180° of Fahrenheit's thermometer. To
make a solution of Indian rubber of about the consistence
of tar, when at a temperature of 60° I put about one pound
of the Indian rubber into one gallon of the essential oil of
turpentine. I mention this only by way of example, and
not as the definite proportions which I uniformly adhere
to, for I vary these according to the nature and value of
the object to which the composition is to be applied. If
the composition is to be made of pitch and of a solution of
Indian rubber, I mix them together, by submitting them
to such a degree of heat as is just sufficient to melt and
keep the pitch in a fluid state, and by stirring it until the
union is complete. Tar, so long as it is sufficiently fluid
(and it generally is so except in very cold weather), mixes
very well with the solution of Indian rubber without being
heated. In making a composition of both pitch and tar
and the solution of Indian rubber, I first melt the pitch,
and mix it with tar in the required proportion, and then
add the solution of Indian rubber. In preparing a compound to be applied to wooden buildings or wooden fences,
&c., by paying or smearing the same therewith, I have
found the following proportions to answer very well;
namely, to one gallon of the essential oil I put one pound
and a half of Indian rubber, mixing with it eight or nine
pounds of tar. If the compound is to be employed in
making ropes, or for covering canvas, &c., I use a solution
of one pound and a half of Indian rubber in one gallon of
essential oil, mixing with it about one pound of pitch, and
six or seven pounds of tar. If the compound is to be used
in paying the bottoms of ships, or to be employed in preparing sheathing for ships, or paper or other substances

P 3

for covering the bottoms of ships or the roofs of houses or other buildings, I dissolve three pounds of India rubber in one gallon of essential oil, and mix it with six pounds of pitch. And, lastly, with regard to the proportions or relative quantities of the materials to be employed, namely, of the essential oil, of the caoutchouc or Indian rubber, of the pitch, of the tar, or of pitch and tar mixed, as have hereinbefore been described, it is evident that such proportions or quantities may be infinitely varied at the discretion of the operator ; and I have only to remark that, in making any of the compounds hereinbefore mentioned, that the elasticity, toughness, and durability of such compound will be increased by increasing the quantity of Indian rubber. I would here also observe that I prefer making use of the best Stockholm tar and Stockholm pitch. Any of the hereinbefore-mentioned compounds, when sufficiently thin and fluid, may be laid on cold with a brush ; those which are thicker require to be moderately warmed, so as to produce the required fluidity : they may then be spread upon wood, paper, canvas, &c., with a brush or trowel, or by any other of the well-known methods for spreading or laying on compositions of a similar nature. In using the trowel, or any other instrument of a similar kind, to spread the composition, I occasionally dip it in water, to prevent the adhesion of the composition to the trowel or instrument. The composition, when employed in making ropes, is applied in the same manner as is commonly practised in making tarred ropes. In all the before-mentioned processes I take care to avoid exposing the composition to a greater degree of heat than about 180° of Fahrenheit's thermometer. Lastly, I do declare that, although I have hereinbefore described the method I employ for making a solution of caoutchouc, or Indian rubber, in essential oil of turpentine, or in other essential oil, I lay no claim to the method of making such solution as forming any part of my invention, the same having been heretofore known and used, but that my invention consists in the admixture of caoutchouc and its solvent with tar, or with pitch, or with both tar and pitch, so as to form a compound possessing the properties of being less soluble in water, of being more tough, of being more elastic, and of being more durable when exposed in the open air, or under water, than tar or pitch, or any mixture of tar and pitch (without the aforesaid admixture) possess when similarly exposed. — In witness, &c.

THOMAS HANCOCK.

Enrolled September 17. 1823.

LEATHER BY LIQUID.

Specification of Patent granted to THOMAS HANCOCK, *Stoke Newington, Middlesex, Esquire, for Improvements in the Method of Making or Manufacturing an Article which may be, in many instances, substituted for Leather, and be applied to various other useful purposes.* — Dated Nov. 29. 1824.

THE nature of my invention consists in combining together the fibres or filaments of various matters, such as flax, hemp, cotton, wool, hair, or other matters of the like flexible nature, by saturating them, in connexion or in contact with each other, with a liquid which, when partially evaporated, becomes a flexible and adhesive substance; such fibres being previously arranged or disposed, as to shape and dimensions, according to the purpose to which they are afterwards to be applied, so as to produce a uniform combination of the fibres and the substance, or in such a manner as that every individual fibre may be so surrounded with the said substance as that the whole of the fibres composing the mass, when united by the substance, may form a compound or article somewhat resembling leather, and which said compound substance or article may in many cases be substituted for leather, more or less advantageously; namely, for harness, straps, belts, accoutrements, boots, shoes, flexible pipes, air-tight bags, and a variety of other things, which

have heretofore been made of leather; and the said substance or article may also be applied to various other useful purposes, such as in parts of wearing apparel which it
may be desirable to have waterproof, and which commonly
are made of other substances than leather. The article may
also be applied for the roofs of verandahs, awnings, tent
coverings, and to other similar purposes. I shall now proceed to describe the method of making the said article, and
the manner in which my said invention is to be performed;
and in order the more clearly to describe the same, I will
premise that the choice of the particular kind of fibres to
be used, and the manner of preparing the said fibres, will
depend upon the use or purpose to which the article is to
be applied when made. Many of the substances I have mentioned, such as hair, wool, cotton, flax, and others, having
short fibres or filaments, are capable of being carded in the
same manner as is commonly practised in carding wool and
cotton for making yarn or thread; I therefore employ the
carding machine (a machine well known and in common
use, and therefore needing no further description) to card
the aforesaid fibres, and those which are capable of being
felted may have that operation performed. And by this
means I am enabled to obtain a layer of fibres of uniform
thickness, and of any convenient dimensions as to length
and width. I take one or more of such layers of fibres,
according to the required thickness of the article when made,
and spread it or them upon a flat board. When the thickness of the article to be made requires more than one of
such layers of fibres, I spread them out upon a flat board,
as before mentioned, layer upon layer, until the number is
completed; I then sprinkle the whole with cold or warm
water (but the latter answers best), at the same time I press
the layers together, until the whole becomes uniformly
wetted. In this state they are placed between two flat boards,
or flat plates, of any suitable metal, and exposed to strong
pressure in a screw or other press, or by passing the said

boards and plates, together with the layers of fibres between them, through or between a pair of rollers. By this means the whole of the fibres become more uniformly wetted, and the superfluous water is expelled, or got rid of. The said layers having been thus wetted are now ready, or in a proper state to be saturated with the liquid which forms the flexible and adhesive substance hereinbefore mentioned. At the time of applying the same it is in a liquid state, and I saturate the layer or layers of fibres therewith, by first pouring a proper quantity of it upon the surface of the said layer or layers; and I spread the same over the said layer or layers with a smooth spatula, or other suitable instrument, made of wood, or other proper material, and by gently pressing the said layer or layers of fibres with the spatula, or other suitable instrument, the liquid is made to sink into, and mix with, and pervade the whole mass of fibres throughout the said layer or layers.   I also perform the said operation of saturating in a more convenient manner, by placing the said layer or layers of fibres in a shallow trough, with a flat and level bottom, and by this means a greater quantity of the said liquid may be poured over the layers, as the sides of the trough will prevent its running to waste; and the fibres are thus more easily saturated, by dabbing or pressing them with the hand or a spatula.   A trough may also be advantageously used for placing the layer or layers of fibres in when they are to be wetted with water, as before described: when the layer or layers of fibres have been sufficiently saturated with the liquid, I place them upon a flat board in an inclining position, and in order to squeeze or press out any excess of the liquid, I pass a wooden roller over the surface of the layer or layers of fibres, at the same time applying a sufficient force by hand, or otherwise, to the roller, to force or squeeze out the liquid : this operation may be carried on by exposing the layers to a strong mechanical pressure, if necessary.   The process having been carried on

so far, the layer or layers of fibres now require to be dried.
I therefore now place the said layer or layers of fibres in a
room heated to eighty or ninety degrees of temperature, and
and allow them to remain there till they are nearly dry, or
until the liquid becomes viscous, glutinous, and adhesive.
The layer or layers of fibres having been thus dried, I expose them again to a strong pressure, by which the whole
of the fibres are brought into closer contact with each other
and with the interposed substance, so as that the said fibres
are made to adhere firmly to each other. If on exposing
the layer or layers of fibres to pressure for the first time
after being dried, I perceive that any water or fluid matter
exudes, or is driven out by the pressure, I conclude that
the first drying has not been continued long enough; I
therefore place the layer or layers of fibres in the warm
room a second time, and afterwards subject the said layer
or layers to a second pressure. It is, in some cases, requisite to give a smooth surface to the article, and this I
effect by polishing the surfaces of the plates between which
the dried layers are pressed for the last time. The liquid
I use for the purpose hereinbefore mentioned is brought
into this country, and said to be the juice obtained from
certain trees which grow in several parts of South America,
the East Indies, and other places abroad. It is stated in
Mr. William Nicholson's translation of Fourcroy's " General System of Chemical Knowledge," that this juice is obtained in South America from a tree called the hevæa. The
juice or liquid I have made use of was obtained from South
America; and from my own experience I find that the
said juice or liquid, when exposed to the open air in the
sun, or in a warm room, becomes inspissated, or dried, and
then forms a substance exactly resembling, and which I
believe to be identically of the same nature and to possess
the same properties with the substance well known by the
names of caoutchouc, or Indian rubber, or elastic gum, and

it is employed abroad for that purpose. Its colour and consistence very much resemble cream. In manufacturing any article where the colour is not an object, I employ this liquid in its natural state, without any previous preparation, excepting that of freeing it from ligneous or other substances by straining it through a sieve or open cloth. But if the article is intended to be of a light or delicate colour, it is necessary to free the liquid from the colouring matter combined with it, and for this purpose I put it into a glass bottle, or other vessel of an appropriate size, and add to it three or four times its bulk of clean water. I then stop or cork the mouth of the bottle or vessel so as to prevent evaporation, and after well shaking the liquid and water together, I allow it to stand undisturbed till I perceive the whole of the water has subsided to the bottom of the vessel, which it does in a few hours, and leaves the liquid floating at the top. The vessel which I employ for this purpose has a hole through its side, at or near the bottom, and through this hole, which is provided with a cork or stopper, I draw off the water after it has completely separated from the said liquid above. This operation of washing or cleansing the liquid I repeat as often as I find necessary for rendering it sufficiently clear and colourless. As the method of manufacturing the article intended to be employed as a substitute for leather, which I have hereinbefore described, applies only to the formation of pieces of uniform texture and strength in every direction, I will now proceed to describe the method I employ in manufacturing any article such as a strap or band, or such as require the greatest strength in one direction only. In such cases I make use of longer fibres, as the long wools, hemp, flax, &c.; and instead of subjecting the said fibres to the carding operation, I cause them to be combed or hackled, and thus lay the greatest number of the said fibres nearly parallel with each other. In the operation of combing wool or hackling flax it necessarily happens that you

obtain a quantity or bundle of long and short fibres, mixed
together in such a manner as to cause the bundle to contain more fibres, and therefore to be thicker, in the middle
than at the ends; or, in other words, the bundle tapers off
from the middle towards the ends. In order, therefore, to
obtain a layer of such fibres of uniform thickness, of a
proper length and width required for the strap I intend to
make, and ultimately to produce a strap of uniform
strength, or nearly so, throughout its length, I first make
a wooden trough of the required length and breadth, with
a flat bottom. In this trough I place in succession small
bundles of such fibres as are proper for the purpose, taking
care that the end of each consecutive bundle shall be laid
upon, or over, the middle of that which preceded it; and
thus filling the trough from side to side, and from end to
end, I obtain a uniform mixture or splicing of the fibres
one with another. The layer being thus completed, I
proceed to wet it, saturate it with the liquid, to dry and
press it in a similar manner to that hereinbefore described.
But although straps and bands require the greatest strength
in the direction of their length, I also find it necessary to
strengthen them in the lateral direction, by interposing
between the layers of longitudinal fibres several layers of
shorter fibres, lying across the strap. If great stiffness,
solidity, and firmness be required in the article to be made,
I saturate the layer, or layers, of fibres with the liquid, as
hereinbefore described, and afterwards press it very lightly
with the roller, in order that a considerable quantity of
the substance may remain incorporated with the layer of
fibres, because the comparative solidity, stiffness, and firmness of the article will depend upon the quantity of the
elastic and adhesive substance remaining in combination
with the fibres, as well as upon the degree of pressure to
which the layers are subjected when dried; and if after
saturation a great portion of the liquid be pressed out, the
softer and more flexible will be the article when finished.

Consequently, when I intend to manufacture an article to
be substituted for the softer kinds of leather, I expose the
layer of fibres, after being saturated with the liquid, to a
greater degree of pressure, so as to drive out a greater
proportion of the liquid.   In making an article of the kind
last mentioned I find it convenient, before I use the liquid,
to mix with it about one fourth of its bulk of water.   After
adding the water to the liquid, they must be shaken together until a uniform mixture is produced, and it must
be used while in this state of mixture.   From what I have
hereinbefore stated, and from the nature of the process,
any competent manufacturer will be led to make a proper
choice of the fineness, coarseness, and kind of fibre to be
used for the various kinds and qualities of the articles intended to be made.   It is necessary for me to mention
here, that when I make use of the liquid in a washed or
purified state, I take care to place the layers of fibres to
be saturated therewith in a trough made of some of the
whiter woods, such as sycamore or American pine, or such
as will communicate no tinge or stain to the liquid or fibres ;
and I also use the same precaution in pressing out the
superfluous quantity of liquid, by effecting the pressure
between boards of the same kind.   And I have also to state
that the liquid acts upon, or is acted upon, by iron, copper,
and brass, and even tin in a slight degree, and that these
metals communicate a stain or tinge, more or less, to the
liquid in contact with them.   But for pressing the layers of
fibres after being saturated and dried, metal plates may
safely be used, and are necessary to give a smooth or glossy
surface.   It will often happen, on account of the great
variety of purposes to which the article may be applied,
that it may, or will, require a combination of the fibres of
several substances to be mixed together, and formed into
one layer, or the combination of two or more layers, in
which the fibres of one layer may be of a different substance from the fibres of another layer: for instance, in
making the article as a fit substitute for leather harness,

where it would be desirable to combine a neat, smooth surface, to great strength, I interpose one or more layers of flax or hemp between two layers of cotton; and other combinations may be made, according to the properties which it may be desired to give to the finished article, or as economy may dictate.—In witness, &c.

THOMAS HANCOCK.

Enrolled May 28. 1825.

LEATHER BY SOLUTION.

Specification of Patent granted to THOMAS HANCOCK, *Stoke
Newington, Middlesex, Esquire, for a New or Improved
Manufacture, which may, in many instances, be used as a
Substitute for Leather, and otherwise.*—Dated March 15.
1825.

MY said invention consists in filling, saturating, and combining various fibrous substances, in their manufactured
and unmanufactured state, with a composition, which leaves
to the fibres sufficient flexibility, and at the same time
unites and consolidates them into one mass, thereby increasing their strength and durability, and producing by
these means a manufacture which may be, in many instances, substituted for leather, and be applied to other useful purposes; such as soles for shoes and boots, hose, pipes,
pails, and other articles which have heretofore been made of
leather ; and also to other useful purposes, such as the roofs
of verandahs, corn and flour-sacks, packing-cloths, and tar-
paulins.   The fibrous substances I employ in this manufacture are wool, cotton, hair, silk, flax, hemp — carded,
combed, or hackled, and combined with the same substances,
woven and manufactured.   As the same process is applicable to all the combinations, it will be necessary to describe

the method I pursue in one case only, as any variation may be made in arranging the different substances at the discretion of the operator. I take a piece of cotton cloth, of any convenient size, and strain it on a board, and spread over it, with a spatula or other convenient instrument, a full coating of one of the compounds to be hereinafter described. I then spread on, or over, the compound a layer of carded cotton, somewhat similar to the article known by the name of wadding, spreading over this again another piece of cotton cloth, prepared as the first. I then submit the whole to sufficient pressure between boards, or plates of metal, either passing them through or between rollers, or otherwise, to force the composition quite through the layer of carded cotton; I then carefully remove it from the boards, or plates, and leave it to dry, either in the open air or in a warm room heated to 80° or 90° of temperature, and proceed to make others in the same manner. When I perceive that they are nearly or quite dry, I again submit them to the press; or if one of these strata is not sufficient to make up the thickness I require, I put two, three, four, or more together, spreading the said compound on the surfaces again if necessary, and increasing the pressure. After they have been in the press some hours they may again be exposed to the air, or returned to the warm room to complete the drying, and if necessary pressed again. When I wish to have the carded cotton for either, or both surfaces, I carefully separate, at the end of two or more pressings, the last layer or layers of cloth from the cotton below it; soon after I take it out of the press, as it will then separate, and proceed as before described. In this manner I introduce into this manufacture hair, wool, silk, hemp, and the like, or any mixture of these fibrous substances, or any or all of them, mixed with chopped hemp or tow, and carded together; or I hackle or comb hemp or flax, and lay the fibres parallel with each other, and combine any intermixture of these

Q

different materials with the different kinds of manufactured wool, silk, linen, cotton, and the like, according to the purpose to which the article is to be applied, or as economy may dictate. For soles of boots and shoes I prefer wool, hair, and cotton, in about equal proportions. For hose, pipes, pails, and accoutrements, chopped hemp, tow, or cotton. I prefer the woven materials to be made of wool or cotton, and these of an open, loose, coarse texture, excepting where it is intended for a finer surface; in such cases I choose the fabric of a finer quality. If the article is required to have a smooth surface, I produce it by using polished metal plates the last time the article is pressed. I make the compound, or compounds, with which I unite or combine the said substances, as follows. (No. 1.) I take two pounds of caoutchouc, dissolved in one gallon of equal parts of oil of turpentine and highly rectified coal-tar oil, six ounces of black resin, two pounds of strong glue size, and one pound of ochre, powdered pumice, or whiting, and mix the whole together; or (No. 2.) one pound and a half of caoutchouc, dissolved as before stated, one pound of strong glue size. I melt and mix the resin and size in a water or steam bath, and then add the other ingredients, stirring the whole until it is mixed throughout. The solution of the caoutchouc is expedited by a water or steam bath, and the undissolved portions may be separated by straining it through a fine wire or other sieve. The mixture No. 1. is applicable to articles where stiffness and cheapness are required; No. 2. is preferable where pliancy and strength are more required. But I think it proper here to state that the proportions above mentioned may be varied according to the different applications of the article to be manufactured. If varied qualities of stiffness or cheapness should be desired, the proportion of size and whiting may be increased till they make up one-third of the mass: if flexibility be required, the quantity of dissolved caoutchouc in the compound No. 2. may be in-

creased, and especially where great strength and pliancy
are required.   This last is also preferable for articles
that are to be much exposed to the weather. — In witness, &c.

THOMAS HANCOCK.

Enrolled September 14. 1825.

ROPES AND CORDAGE.

Specification of Patent granted to THOMAS HANCOCK, *Stoke Newington, Middlesex, Esquire, for an Improvement or Improvements in the Preparation or in the Process of Making or Manufacturing of Ropes or Cordage, and other Articles from Hemp, Flax, and other Fibrous Substances.* — Dated March 15. 1825.

THE nature of my said improvement, or improvements, in the preparation, or in the process of making or manufacturing of ropes or cordage, and other articles from hemp, flax, and other fibrous substances, consists in the mixing, or covering, the said fibrous substances when they are formed into yarns, strands, ropes, or cordage, or woollen, cotton, or silk threads, with a liquid which I shall hereinafter more particularly describe, so as to render the said substances, when formed into ropes, cordage, or other articles, less liable to injury from air or moisture, and thereby increasing their durability. The liquid I use for the purpose hereinbefore mentioned is brought into this country, and is said to be the juice obtained from certain trees which grow in several parts of South America, the East Indies, and other places abroad. It is stated in Mr.

William Nicholson's translation of Fourcroy's "General System of Chemical Knowledge," that this juice is obtained in South America from a tree called the hevæa. The juice, or liquid, I have made use of was obtained from South America; and from my own experience I find that the said juice, or liquid, when exposed to the open air in the sun, or in a warm room, becomes inspissated, or dried, and then forms a substance exactly resembling, and which I believe to be identically of the same nature, and to possess the same properties with, the substance manufactured into bottles and other articles, and well known by the names of caoutchouc, or Indian rubber, or elastic gum. In colour and consistence the said juice very much resembles cream. The application of the said liquid to the aforesaid fibrous substances, when formed into yarns, strands, ropes or cordage, or woollen, or cotton, or silk threads, is easily effected; and to this end I employ a process (as being a convenient one) exactly similar to that employed by ropemakers in making tarred ropes, or in saturating the strands of the rope with tar, and which process is well known to persons employed in rope-making, with this difference only, that I employ the said liquid instead of the tar; but the said liquid is not heated, and the strands, yarns, or threads must be separated whilst in their wet state, or they will become more or less united. If it is wished to increase the quantity of the liquid on the yarns, threads, or ropes, I effect this by repeated coatings, until the required thickness is obtained, laying it on always just before the previous coating is quite dry, and keeping them separated. The substances having been thus saturated, or coated, with the liquid, I then place them in a warm room, or in the open air, until the evaporable part of the liquid has been completely dissipated. I lastly declare, that my invention consists in the application of the liquid hereinbefore described to ropes or cordage, and other articles, such as woollen, yarn, silk, and cotton

Q 3

thread, or to ropes, cordage, strands, yarns, or threads made from other fibrous substances, so as to render the ropes, cordage, strands, yarns, or threads, less liable to injury from exposure to air or moisture. — In witness, &c.

THOMAS HANCOCK.

Enrolled September 14. 1825.

ORNAMENTS, ETC., BY LIQUID.

Specification of Patent granted to THOMAS HANCOCK, *Stoke Newington, Middlesex, Esquire, for Improvements in the Manufacture of certain Articles of Dress or Wearing Apparel, Fancy Ornaments, and Figures, and in the Method of rendering certain Manufactures and Substances in a degree, or entirely, impervious to Air and Water, and of protecting certain Manufactures and Substances from being injured by Air, Water, or Moisture.* —Dated August 5. 1830.

THE improvements in the manufacture of certain articles of dress and wearing apparel, fancy ornaments, and figures, consist in making them of a certain fibrous material, as hereinafter described, and coating them, where necessary, with a certain liquid composition, or varnish, as hereinafter also described.

The principal ingredient used in the improvements for which this patent is granted is liquid caoutchouc, which is obtained from South America, the East Indies, and other places, and when dried forms the substance called Indian rubber, and is more particularly described in the specification enrolled in the Petty Bag Office in the Court of Chancery, under a patent for an invention granted to me, dated the 29th day of November, in the year 1824, to which I hereby refer. The fibrous compound, or material, is made by mixing hair, wool, cotton, or other fibrous substance, with liquid caoutchouc, to which should be added certain substances, according to the intended object, such as whiting, ochre, brickdust, emery-powder, or other such substances. The following proportions may be advantageously used : 10 lbs. of liquid caoutchouc, 10 oz. of whiting, 10 oz. of Oxford ochre, and 10 oz. of hair, wool, cotton, or other similar fibrous substances. The whiting and ochre should be first mixed with a very small quantity of water ; and the hair, wool, or cotton, or fibrous substance, is generally better cut very short, averaging about one-eighth of an inch in the staple. The whole should be well mixed together. Sheets may be made of this compound by spreading some of it on a flat surface, such as a stone, or on plaster cast, and leaving it to dry, either in the open air or in a warm room; when dry it may be made more compact and firm by pressure. This may be effected by passing it between rollers, or pressing it in a common screw-press between plates ; and the operation will be facilitated by the temperature of the plates and rollers being kept while used at about blood-heat. These ingredients may be altered or varied very considerably, according to the purposes to which it is intended to apply the material when made. If intended for purposes where sewing is required, the quantity of fibrous ingredients should be increased, and the fibres used somewhat longer. If greater stiffness is required, a small quantity of glue-size, thin starch, or paste, or a thin solution of gum-arabic, may be added. For some purposes the wet pulp of rags, such as is used by papermakers, may be used instead of the before-mentioned fibrous substances. The colour may be varied by adding a little lamp-black, chrome yellow, blue verditer, Venetian red, or other colouring matter. With these, sheets, gaiters, goloshes, boots, shoes, hats, caps, and other articles of dress and wearing apparel, may be manufactured in the usual manner. Hats, shoes, caps, and other such articles of dress and wearing apparel which may be moulded, and also fancy ornaments and figures, may be manufactured by preparing moulds of plaster of Paris, or other substances. The fibrous compound should be poured into the mould, and allowed to stand a few minutes, and all the compound that will then run out should be poured from it, and the mould be allowed to stand till that which remains is nearly dry. The mould should be again filled, and so from time to time till the proper thickness is obtained for the article or purpose required. In forming large figures, such as the human figure, anatomical subjects or busts, that part of the figure which forms the aperture by which the compound is poured into the mould, should be formed into a neck, in which a valve or stop-cock should be inserted, by which sufficient air may be introduced into the figure when it is taken out of the mould to fill out its form. Or the aperture may be made larger, so that the figure may be filled with bran, fine sawdust, or any other such material; or it may be filled, before it is taken out of the mould, with glue dissolved in equal parts of water and treacle, by boiling. For shoes, hats, caps, and other articles requiring strength, the quantity of fibrous matter may be increased, and the thickness of any part may be increased by pouring into the mould a small quantity of the compound, and allowing it to cover that part only, and repeat the operation as before stated. The soles of shoes or boots may be thickened in the same manner, or they may be formed of the sheets before described, and cemented on to the other part of the shoes or boots by a fresh application of the compound, or liquid caoutchouc. In forming the sheets for soles of shoes, or other articles requiring great firmness, it is better to substitute for the ochre and whiting harder substances, such as brickdust or fine emery. If from any cause so great a quantity of fibrous or other ingredients should have been mixed with the liquid caoutchouc, that the compound formed should be too porous for the designed purpose, coatings of the liquid caoutchouc may he applied in sufficient quantities to bring it to the desired closeness.   The improvements in the method of rendering certain manufactures and substances in a degree, or entirely, impervious to air and water, and of protecting certain manufactures and substances from being injured by air, water, and moisture, consist in the application, in the manner hereinafter described, of liquid caoutchouc, having mixed with it as much of the colouring matters hereinbefore mentioned, or others, such as extract of cochineal, logwood, indigo, lake liquor, or red or black ink, as should be thought desirable to give it any required tint, well stirring and mixing it together, and using no more of the colouring matter than is necessary.   The colouring substances, when solid, must be first ground very fine in water, and the whole composition well stirred and mixed together ; but no more of the colouring matter must be used than necessary to give the colour required.   If a light colour should be required, and the liquid caoutchouc should be too dark, it may be washed by adding to it pure water, in a proper vessel, and shaking it well together.   It should be then left undisturbed ; the pure caoutchouc will separate and float on the top, and the one may be drawn off from the other.   Most of the colours require that this last-mentioned composition or varnish should be used immediately it is mixed ; and where the article is intended to be used under any exposure to the rays of the sun, the colour should be black or dark, as otherwise caoutchouc when so exposed is liable to soften and crack.   When the sheets before described are required (from the use to which it is intended they should be applied) to be exposed to the sun's rays, or to be rendered more completely waterproof, the coloured composition or varnish should be applied to them after they have been rolled or pressed, spreading it on them as uniformly as possible with a spatula, or other convenient instrument; or by diluting it with water, and laying it on with a brush or other suitable instrument, and when nearly dry repeating the coating either over the whole or a part, as the case may require. When a fine surface is required, the sheets should be again pressed, when dry, between smooth or polished rollers or plates. The various other manufactured articles may be made waterproof by applying the varnish in the same manner as to the sheets, except that in some cases they cannot be conveniently subjected to pressure. Where articles of the fibrous compound which are formed in moulds require to be protected from moisture or air, and it is wished to give them a fine surface, the coloured composition or varnish should be first poured into the moulds and allowed to dry, and then as many coatings of the fibrous compound, or of the liquid caoutchouc, added as may be necessary. To render articles made of linen, cotton, woollen, silk. leather, and also packing-cases, tanks, and other articles of wood or metal, impervious to water and air, more or less, as may be required, they are to be coated on one or both sides in such thickness as may be necessary with the coloured composition or varnish ; and in the same manner, by coatings of the varnish, substances such as metals, paper, plaster casts, walls and the interior of rooms may be protected from being injured by air, water, or moisture. If one part of any article is not required to be so perfectly impervious to water as another, the coloured composition or varnish may be diluted by mixing with it one third of its weight of water, or more or less, according to the degree of protection required, and giving that part only one or two coatings. A boot or shoe, for instance, made of woollen or other cloth, felt, or leather, may first have applied one or two coatings of the diluted varnish; and then, beginning at the sole, as many coatings of the thicker varnish as shall be necessary, according to the substance of which the boot or shoe is formed, may be applied as high as it may be wished that the boot or shoe should be more perfectly impervious. If the article is required to be impervious to water throughout, such as fishing-boots, or stockings, gloves, hose, pipe, or bags to contain liquid, they may be filled with the thicker varnish, which should be allowed to remain long enough within the article to saturate the material of which it is composed. The varnish should then be poured out, its sides being kept from collapsing until the varnish is dry. The article may then be immersed in the varnish, so as to saturate the outside, and the external coating allowed to dry; and this proceeding may be continued from time to time until the article is waterproof. The operation may be carried on and completed on one side only, either within or without. Articles intended to contain air, such as beds, pillows, or cushions, may be formed of linen, cotton, silk, or other suitable material. They must be sewed strongly together, and made with the necessary partitions, so as to preserve their form, and an aperture or apertures should be made by which they may be inflated. They should be immersed in the liquid caoutchouc, and then withdrawn, or it should be spread over them; and this should be repeated as each coating becomes dry, or nearly so, until the material is well saturated. The article should then be finished with the coloured varnish, adding as many coatings as may be necessary to render them air-tight. Cloths of all kinds, leather or other materials for great-coats, cloaks, gaiters, travelling bags, portmanteaus, wrappers, flour-sacks, cart-covers, and other purposes required to be rendered waterproof by the varnish, may be strained on frames, and the varnish applied in repeated coatings until they become impermeable to moisture. If it is wished to form any impression of figures, letters, or ornaments, on the surface of any of the articles for which the fibrous compound or varnish is used, it may be done with a stamp appropriate to the purpose before the composition is quite dry, when a slight pressure is sufficient. Although certain proportions of materials have been hereinbefore stated as those which may be advantageously used, it is evident that such proportions may be considerably varied, according to the discretion of the operator and the quality of the liquid caoutchouc; but it is to be observed that the elasticity and toughness of the compound is increased with the proportion of the caoutchouc.

Lastly, I hereby state that I only claim as the subject of this patent the manufacture of various articles as hereinbefore mentioned, of the combination or compound of liquid caoutchouc with fibrous and other matters, as hereinbefore described, and the application of the coloured composition or varnish to the various purposes as hereinbefore also described. — In witness, &c.

THOMAS HANCOCK.

Enrolled October 5. 1830,

EXPANDING CUSHIONS.

Specification of Patent granted to THOMAS HANCOCK, *Stoke Newington, Middlesex, Esquire, for an Improvement or Improvements in Air-Beds, Cushions, and other Articles manufactured from Caoutchouc or Indian Rubber, or of Cloth or other flexible material, coated or lined with Caoutchouc or Indian Rubber.* — Dated June 4. 1835.

THE nature of the said invention consists in the application of strips of Indian rubber to caoutchouc, or to India rubber cloth or other material of which the said air-beds, cushions, or other articles intended to be inflated with air are made, so as to contract or gather up the cloth, India rubber, or other material, in order -to cause or increase elasticity therein.   And the manner in which the same invention is to be performed is as follows (that is to say): If the material to be used in making the said air-beds, cushions, or other articles, be India rubber only, a flat piece of India rubber should obtained sufficient to form one side, or half of the article.   For this purpose India rubber in sheets may be conveniently used, such as are sold in the shops; and if the sheets are not sufficiently large, two or more sheets may be joined by the well-known process of cutting their edges straight, and, after making them quite warm,

bringing the edges into contact, and pressing them together. The piece or sheet of India rubber should be then
laid upon a flat board or surface, of somewhat larger
dimensions than the sheet of India rubber itself. Some
of the best India rubber should then be procured, and that
known by the name of Bottle India rubber, being of good
quality, will answer the purpose. This should be cut
into thin strips, which may be called contractile strips.
For cushions and other articles requiring a similar elasticity, the strips may be one sixteenth of an inch thick,
and three eighths of an inch wide; for beds and articles
requiring a similar elasticity one quarter of an inch thick,
and half an inch wide. The particular dimensions of these
strips are not essential. If the thickness be increased, and
the width diminished in the same proportion, the effect will
be nearly the same; and by diminishing the sum of the
dimensions the force of contraction will be diminished, and
by increasing the dimensions it will be increased. These
strips should then be heated to a temperature of from 150°
to 200°, by which means they become less liable to crack
or break when extended or stretched; this may be easily
accomplished by putting the strips into water heated to that
temperature. This temperature is not essential, as the
effect may be accomplished, but not so well, without the
high temperature mentioned. The strips when dry, and cut
of the proper length, should be extended or stretched almost
as much as they can bear without breaking, and should be
laid across the flat piece or sheet of India rubber, and
fastened in their state of extension, which may be easily
effected by a nail or tack driven through each end, carried
beyond the flat piece or sheet of India rubber and the upper
surface of the board into the sides or edges, or to the
underside of the board on which it may be placed. The
number of contractile strips must vary according to the
elasticity or contraction required. For cushions and articles
requiring similar elasticity the strips may be placed from

three to four inches apart, and for beds and articles requiring similar elasticity from six to eight inches apart, and for other articles in proportion. The distance and arrangement of the strips may be varied as the manufacturer shall choose to vary or increase or diminish their contractile effect. The contractile strips being thus placed upon the said flat piece or sheet of India rubber, they should be pressed down upon it with a moderate force until they become united to it, which will be effected in a short period. The piece or sheet of India rubber must then be coated with thin glue, size, and whiting, leaving uncoated, for cushions and similar articles, about half an inch round the edge, and half an inch across where each strip is placed (and of which half inch the strip will be better in the middle); except that across or over each strip one or more spaces or lines, of about an inch broad, must be coated with the whiting, so as to form a coated or protected communication from space to space between each strip. For beds and similar articles the space uncoated may be an inch broad, instead of half an inch. To complete the article another piece or sheet of India rubber similar to the first should be prepared in the same manner, with the same number of strips, placed in the same relative situations, and coated in the same manner. The two pieces or sheets of India rubber, while remaining upon the boards, must be placed one upon the other, accurately bringing into contact the coated and uncoated parts respectively of each sheet, and the whole submitted to moderate pressure by weights or otherwise. After remaining about an hour, the parts which were uncoated will have become united, and those parts should be further well rubbed down with the hand; the parts which were coated will remain disunited, and form receptacles for the air. The article may then be taken from the board, and the ends of the strips cut off. If the India rubber used is not perfectly clean, or has been long exposed to the air, or from any other cause does not readily unite, the parts upon which the contractile strips are to be laid, and all the parts intended to be united should be coated with a solution of India rubber, taking care that such coatings are nearly dry before the parts are brought into contact. Instead of thin glue, size, or whiting, other similar preparations may be used, or the spaces intended to remain disunited may be lined with cloth or thin kid, or other similar leather; such lining being made to adhere to the India rubber by coating it with a varnish or cement formed of a solution of India rubber, and applied when the coating is nearly dry. Articles lined in this manner will be much stronger, but their flexibility will be lessened. In some cases, where greater ease is intended, free admission of air throughout the whole cushion or other article may be insured by keeping open the passages across the contractile strips by a tube formed of coiled wire, and covered with sheet India rubber, put into the said passages. For some articles, such as pillows, when it is not necessary or advantageous that the two opposite sides should be united, except at the edges, the whole pieces or sheets, except the edges, should be coated with the whiting, or other preparation, when upon the boards, before the pressure to unite the edges is applied. If it be intended that the article should remain permanently inflated, a small taper tube of metal may be introduced into or through any part of the edges intended to be joined, and air forced in by the mouth or any other convenient means, and when filled sufficiently the tube may be withdrawn, and the opening left by the tube quickly closed by pressure; great care must be taken to prevent wetting the opening. Or if the article is intended to be occasionally inflated, a small cock may be fastened into the united edges, by covering the part to be inserted with sheet India rubber bound tightly round it, and introducing it either at the time the union of the edges is effected, or a slip of stiff paper may be laid, to prevent the union until the other parts are finished, and the slip of paper then with-

R

drawn, and the cock, prepared as before directed, inserted;
and in order to facilitate the introduction of the said cock,
and more effectually insure its perfect junction to the side
of the aperture into which it is inserted, the surface of that
part of the said cock which is to enter the said aperture
should be coated with a solution of India rubber, and the
cock introduced whilst the solution yet remains in the
moist state, or previously to its becoming dry.   In an hour
or two after this operation, or when the said solution becomes
dry, care must be taken to make the junction secure by
pressure all round the inserted parts of the said cock. The
article being completely manufactured, except as to this
last operation, must be exposed for some time to a moderate
heat (and the placing or holding it near a fire will produce
the desired effect), when the strips of India rubber, which
will have been previously in an extended state, will exert
their contractile force, and gradually contract or gather up
the sheet of India rubber, and give the intended elasticity.
Instead of the sheet or flat piece of India rubber, the patent
India-rubber cloth, or leather, or any other flexible material coated or lined with India rubber, may be used.   In
such cases the solution of India rubber may be used as a
cement, where the parts are intended to be united in the
same manner as is hereinbefore directed when the India-
rubber sheet has been long exposed to the air.   Cushions or
other articles may be made exactly in the form and manner
in which they are now made with the patent India-rubber
cloth or leather, only first applying the contractile strips of
India rubber to the inside or varnished side of the India-
rubber cloth or leather, in the manner hereinbefore described; and when the said cushions or beds are completed, applying the same to the fire, or a moderately high
temperature, to produce the contractile action of the strips
in a manner hereinbefore described.   Contractile strips of
India rubber may also be applied to articles already made of
patent India-rubber cloth or leather, or other flexible mate-

rial lined or coated with India rubber. For this purpose the article should be coated with the solution of India rubber where the contractile strip and the ribbon or tape hereinafter mentioned are intended to be applied; and the contractile strips may then be laid across the outside of the said article in the same manner as across the sheets of India rubber hereinbefore described: and the contractile strips should be covered with ribbon or tape, which may be made to adhere by means of the cement of the dissolved India rubber, and then the contractile force of the India rubber may be excited by the fire or heat in manner hereinbefore mentioned. The solution of India rubber is well known, and may be made for the purposes hereinbefore mentioned, by cutting sheet India rubber into shreds, and putting it into essential oil of turpentine, in the proportion of about eighteen or twenty ounces of the India rubber to one gallon of the oil of turpentine, and stirring it occasionally for one or two days, when it is fit for use. The benefit of this patent is not claimed to extend to the manufacture of the solution of India rubber, nor to the method of inflating articles with air by means of tubes or cocks, as hereinbefore described; but it is claimed only to extend to the use or application of strips of India rubber to air-beds, air-cushions, and other articles to be inflated with air, manufactured from caoutchouc or India rubber, or of cloth or other flexible material, coated or lined with India rubber, in order to produce contraction in the manner hereinbefore described. And for this purpose the said use or application is not intended to be limited to the specific dimensions, distances, or directions of the said contractile strips; these may be varied to a considerable extent at the pleasure of the manufacturer. — In witness, &c.

THOMAS HANCOCK.

Enrolled December 4, 1835.

DOUGH WATERPROOFING.

Specification of Patent granted to THOMAS HANCOCK, *Stoke Newington, Middlesex, Esquire, for an Improvement or Improvements in the process of rendering Cloth and other Fabrics partially or entirely impervious to Air and Water by means of Caoutchouc or India Rubber.*—Dated April 18, 1837.

INDIA RUBBER has for some time past been prepared and manufactured into masses for the purpose of being cut into sheets, and for other purposes, and has been called or known by the name of prepared, manufactured, or patent India rubber. A method of preparing or manufacturing the same is shown in the annexed drawing.* Fig. 1 is a cross section, and Fig. 2 is a side-view of the apparatus, with a portion removed in the latter to show the shaft in the interior. The letters refer to the same parts in both these figures. A is a hollow iron cylinder, with closed ends fixed to a frame. B C is a door in the cylinder, about one-third its circumference, and of its whole length. D is a fastening to keep it closed. E is a shaft passing through the cylinder A,

---

* This drawing is not given here, as it has been previously described as the "Masticating machine."

and filling about two-thirds of its internal diameter; this part of the shaft is grooved, so as to form projections about half an inch square along the whole of its length within the cylinder; the ends of this shaft are turned of a suitable size to pass through the ends of the cylinder, and to run upon bearings on the frame B; this shaft may be driven at the rate of about thirty revolutions a minute, — it is not necessary to heat the apparatus, as it speedily becomes heated in use. The India rubber should be cut in rather small pieces and well cleaned and dried, particularly the white cake India rubber; the whole is then to be warmed to about from 100° to 150° Fahrenheit, and passed through rollers, which will also soon become sufficiently warm in use. This operation will, in some measure, unite the pieces into sheets or lengths, and may be repeated two or three times, after which the India rubber is to be put into the cylinder A, and the door fastened. Motion being now given to the shaft, it is communicated to the India rubber; and in the course of half an hour, more or less, according to the speed of the shaft and the quantity of India rubber employed, the combined action of heat and friction, occasioned by the motion and pressure on the India rubber, has the effect of uniting it into one compact mass or roll (as shown at F), of the length of the interior of the cylinder. The perfect uniformity of the whole mass may be ascertained by cutting a thin piece off and holding it up to the light, when, if there are any granular particles to be seen, the operation must be resumed until they disappear. The quantity or charge of India rubber to be operated on at one time may be regulated according to the size of the apparatus and speed of the shaft. By the addition of more India rubber the pressure is increased, and the operation quickened. Increasing the speed has a similar effect, but a little practice will soon enable the workman to ascertain the quantity proper for the apparatus, and the power at his disposal. The length of the cylinder is not material, when

R 3

the mass or roll formed by this apparatus is intended to be rolled into sheets for making the softened India rubber; but when intended for spreading on the cloth without solvent, I prefer that the length of the cylinder, and of course the roll of India rubber produced, should be of the width of the cloth upon which it is to be spread. This method of preparing India rubber has been some time in use, and is not claimed to be protected by this patent. This manufactured or prepared India rubber is well adapted to be used in the processes hereinafter described. A portion of this prepared or manufactured India rubber (and for this purpose about sixteen pounds will be a convenient quantity) must be warmed in a stove heated to about 100° of Fahrenheit, at which temperature it should be passed between metal rollers, heated also to about the same temperature, and so formed into rough sheets of about one-sixteenth of an inch or less in thickness. The ordinary sheet India rubber may be used, but it is more expensive, and less fitted for the purpose. The raw or unprepared India rubber may be used after being passed a sufficient number of times between the heated metal rollers, till it is reduced into sheets; but the prepared or manufactured India rubber I consider the best for the purpose. Heated metal rollers are now so commonly used that no description is necessary. To these sheets about eight pounds of coal oil, or other solvent of India rubber, must be applied, so that the whole surface of the sheets shall be wetted as equally as possible, and so that the eight pounds of the solvent shall be absorbed by the India rubber. This may be effected by passing the sheets through the solvent, placed in a vessel, or by applying the solvent to the surface by brushes, or other suitable means. If it is wished to add any colouring matter, it may be sifted on during, or immediately after, this operation. The sheets so prepared should be left in a covered vessel for ten or twelve hours. The whole should then be mixed and blended together, and the apparatus in which the com-

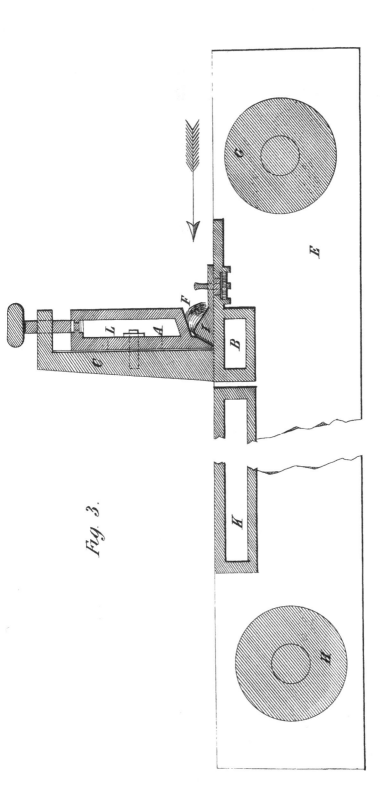

Fig. 3.

Alfred. S. Pexter, Lith. 58, Fleet Street.

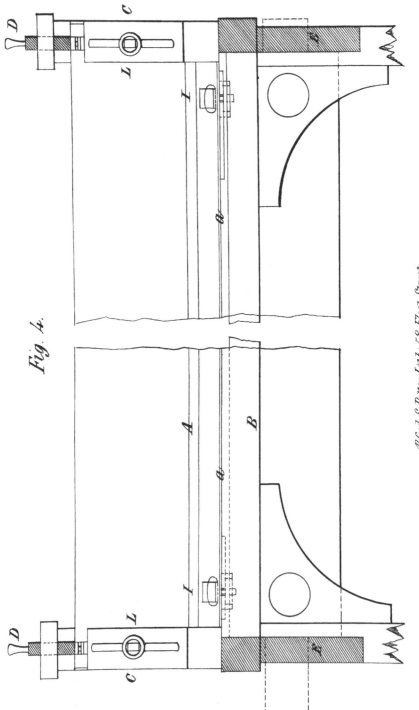

Fig. 4.

Alfred S. Petter, Lith. 58, Fleet Street.

mon India rubber is manufactured, or prepared to be cut into sheets, as hereinbefore described, is well adapted for the purpose. By this process the India rubber will be brought into a state more soft, and less tenacious and elastic than the ordinary India rubber. It will be rather more firm than dough or putty, and have still somewhat of the adhesive or elastic character of India rubber, but it may be compressed and moulded by the hand India rubber thus softened by, and combined with, a solvent, may be spread or applied in manner hereinafter described on any of the fabrics now in general use for such purposes; but should it be required to be applied to any fabric more than usually delicate, and liable to be injured by the force necessary to be used, a somewhat greater proportion of the solvent should be used, according to the delicacy and fragility of such fabric. The India rubber thus softened is to be applied to the fabric to be made waterproof by the spreading machine or instrument delineated in the annexed drawing, Figs. 3 and 4. E is a frame fixed, with suitable bearings, to support the rolling and other machinery attached to it. B is a hollow piece or box of metal fixed firmly in the frame E, which hollow piece I have denominated the bed. That part of the bed B immediately under the lower edge of the spreader A, hereafter described, is elevated a little above the rest of the bed, and is about three or four inches wide, and perfectly straight, flat, and smooth. C C are uprights, firm and strong, intended to hold and guide the spreader A. L L are screws fixed on the uprights C C, passing through grooves in the spreader A, to keep it steadily in its place as it is required to act. The spreader A is a hollow box of metal, brought down at the bottom nearly to an edge, not quite sharp, but somewhat rounded. The bottom line should be perfectly straight, so as accurately to coincide with the elevated part of the bed B ; and by means of the screws D D, it may be fixed firmly at any determined distance therefrom. The length of A and B between the uprights C C, should be rather greater than the width of the cloth intended to be made waterproof, and the spreader A should be screwed down so close to B as just to allow the fabric, covered with a film of India rubber, of any determined thickness, to pass through. In order to prevent the India rubber spreading beyond the width of the cloth, two pieces of metal, fitted to A and B, and working in grooves, and capable of being adjusted by screws, should be attached to B, as shown at I, Figs. 3 and 4. The fabric intended to be made waterproof must then be attached at one end to a tacking-cloth of a suitable material, which must be tacked to and rolled round the roller G, with the fabric to be made waterproof. The other end of the fabric is to be brought up to, and just passed through beneath the spreader A, and there attached to another tacking-cloth of similar material, which latter tacking-cloth should be tacked to the roller H. If the fabric to be made waterproof be of delicate texture, another cloth should be attached to it, so as to line or cover the underside throughout, and prevent any injurious strain, and pass under the spreader with it. The metal spreader A, and the bed B, should be heated to a temperature of from 85° to 100° of Fahrenheit, and the spreader A should be adjusted by means of the screws to the intended thickness of the coating of India rubber. A warm roll of the softened India rubber F is then to be pressed down on the fabric to be made waterproof, and in contact with the heated spreader. A slow motion is then to be given to the fabric by any mechanical power at the roller H, and the fabric drawn beneath the spreader, by which means a smooth, thin coating of the India rubber will be spread upon it, which process may be repeated until the coating of India rubber is of sufficient thickness. The speed at which the fabric should pass beneath the spreader may require to be varied in using softened India rubber of different temperatures, or when formed of different propor-tions of materials; two minutes to the yard will be found generally to answer very well. After the fabric has been drawn beneath the spreader, it must be passed over a heated cylinder, or other heated surface, as shown at K, in Fig. 3, raised to a temperature of from about 100° to 150° Fahrenheit, of sufficient extent to dry the India-rubber film before it is rolled up. It is obvious that the form of the spreading machine may be varied without affecting the principle of the process. Instead of the metal box, B, a revolving hollow cylinder, heated to the same temperature as the spreader A, and adjusted to the proper distance from it, may be used; to which cylinder motion may be given by any ordinary means, and this form of the machine will answer the purpose very well. By means of the spreading machine hereinbefore described, when heated to a sufficient temperature, the prepared or manufactured India rubber hereinbefore first described, without any solvent, may be spread upon the cloth or fabric in the manner hereinbefore described for the softened India rubber. But the prepared India rubber does not adhere with certainty so firmly to the cloth without the temperature of the spreader being greatly raised, which it is desirable to avoid; as in case of any accidental stoppage of the process the India rubber in so thin a film may become in some degree decomposed. Two textures are united by coating each with the India rubber as before described, and passing them with sufficient pressure between iron or wooden rollers, with the coated surfaces in contact. I have now described the method of making cloth and other fabrics waterproof: but I do not claim as any part of my improvements the use of the machine for the preparing of India rubber, delineated in Figs 1 and 2, which has been for some time in use, nor the use of the prepared or manufactured India rubber, as hereinbefore first described, for the purpose of making fabrics waterproof; nor the use of India rubber rendered fluid or semifluid by means of any solvent. But I claim as my invention the use and application of India rubber, softened to the consistency hereinbefore described, by intimately combining therewith a small quantity of solvent, for the purpose of making the said cloth and other fabrics waterproof, in manner hereinbefore described; by which means much less expense is necessary than in the manufacture of waterproof cloth and other fabrics by means of fluid or semifluid India rubber.   And I also claim the method of applying the India rubber softened as hereinbefore mentioned, or the prepared India rubber hereinbefore first mentioned, by the use of the spreading machine hereinbefore described, heated as hereinbefore mentioned, together with the use of the conducting cloth, where the texture of the fabric is thin or delicate, for rendering cloth and other fabrics partially or entirely impervious to air and water. — In witness, &c.

THOMAS HANCOCK.

Enrolled October 17, 1837.

DOUGH SHEETS.

Specification of the Patent granted to THOMAS HANCOCK, *of Goswell Mews, in the County of Middlesex, Patent Waterproof Cloth Manufacturer, for Improvements in the Method of Manufacturing or Preparing Caoutchouc, either alone or in Combination with other Substances.*— Sealed January 23, 1838.

To all to whom these presents shall come, &c. &c.— *Now know ye,* that in compliance with the said proviso, I, the said Thomas Hancock, do hereby describe and ascertain the nature of my said invention, and the manner in which the same is to be performed by the following description thereof (that is to say): —

My improvements in the method of manufacturing caoutchouc consist in forming this substance into sheets and into long uniform slips or threads, in the manner following: I prepare the caoutchouc for forming these sheets in the manner fully described in the specification of a patent granted to me for an improvement or improvements in the process of rendering cloth and other fabrics partially or entirely impervious to air and water, by means of caoutchouc or India rubber, bearing date the 18th day of April,

1837, and to that specification I refer for the particulars of such process.

I will now describe the manner in which I form this prepared caoutchouc into sheets. I take linen, silk, cotton, or other suitable cloth, and saturate or fill the texture with common glue, size, gum, paste, or any other similar substance easily removable by water, and when dry I pass this cloth through a calender, or otherwise smooth the surface of it more or less, as the case may require. I then spread the prepared caoutchouc upon this cloth by means of a machine similar to that described in my said former specification as suitable for this purpose. If one coating is not sufficient, I repeat the operation, and when the coating is dry I immerse the whole in moderately warm water, and let it remain until the gum or size is sufficiently softened to allow the caoutchouc to be separated from the cloth. If the sheets are required of greater thickness than can be conveniently made by successive coatings upon one cloth, I spread the caoutchouc upon two cloths, and unite the two coated sides together before they are quite dry, and I then strip off the cloth from one side, and apply a third coating, which has been spread upon prepared cloth, in the manner before described, on that side, and continue to add fresh coatings in the same manner, until the required thickness is obtained. Instead of filling the cloth with gum or size, one side of it may be covered with paper pasted on, and caoutchouc spread upon it, and the cloth and sheet separated by immersion in water, as before stated. If it is desired to attach the sheets permanently to cloth, leather, &c., I spread a coating or two of the well-known common solution of caoutchouc upon it by way of cement, and when nearly dry unite the sheets thereto by pressure, and then strip off the gummed cloth by immersion in water, as before described. Any suitable pattern or graining may be given to the sheets of caoutchouc by raising figures previously on the prepared cloth,

either by embossing or otherwise. Colouring matter may
be added to the caoutchouc in the manner described in
the aforesaid specification; or the sheets may be coloured
by laying on with a brush any of the common pigments,
such as flake white, vermilion, lamp black, blue verditer,
&c., mixed with a very thin solution of caoutchouc, composed of about ten ounces of manufactured caoutchouc
to one gallon of rectified coal oil, or turpentine. The
sheets may be printed with the same colours by means
of plates, type, blocks, or stencilling, in the manner commonly practised. I also form sheets to be used as tablets
for writing or drawing on, with crayons, French chalk, &c.,
by adding pumice powder, fine emery, or other similar
gritty substances, either in the manner described in my said
former specification, with reference to colouring matters
during the preparation of the caoutchouc, or by coatings of
these substances mixed with a thin solution of caoutchouc
after the sheets are made, as before described. I also form
sheets from the original native and liquid caoutchouc as
imported from South America, by preparing the cloth with
gum or size, as before stated; and I find a convenient
mode of doing this is by attaching two of the gummed
cloths together by means of paste or gum, and then immersing the cloths in the liquid caoutchouc, and allowing
the superfluous liquid to run off. I then hang up the cloth
to dry, and when dry immerse it again in the contrary
direction, and again allow it to dry, and continue the operation until the sheet has acquired the desired thickness; I
then separate the cloth from it by immersion in water, in
the manner before described. The edges must be trimmed
or cut off, if necessary, to allow the water to penetrate
into the cloth. The surface of the cloth may be either
plain or figured, as before described, and colouring matter,
or emery, or pumice powder, introduced into the first or
subsequent coatings. If the sheets are required to be only

of a comparatively small size, or of particular shapes, I form the flat or figured surface upon which I make them of plaster of Paris, and pour on the liquid caoutchouc. If required of considerable thickness, I find it best to pour on a thin coating first, and allow it to dry, and continue to do so until the desired thickness is obtained. If a very smooth surface is wished, I spread it upon plate-glass. In some cases I employ a raised edge of wood or other material as a guide to regulate the required thickness of the sheet, and to prevent the caoutchouc from spreading too far. I also find for some purposes that a coating of native liquid caoutchouc improves the sheets formed of manufactured caoutchouc first described; in such cases I apply it either by dipping the sheets in the liquid, or spreading it on with a spatula, or other convenient instrument. I manufacture the long uniform slips or threads from the native liquid caoutchouc upon cylinders of wood or metal, by turning or otherwise forming a spiral groove of the width and depth of the required slip or thread. I fill this groove by immersing the cylinders in the liquid, and with a straight piece of wood or metal clear the liquid from the projecting parts; and when dry I immerse it again, and continue so to do until the groove is filled, when, by immersing the cylinder for some time in moderately warm water, the slip or thread of caoutchouc may be drawn off, and will be of the length and size of the spiral groove. I also follow the same process of coating a plain cylinder with the liquid, until a uniform substance of the required thickness is obtained; the cylinder is then put into a machine attached to a screw motion, and the slips or threads cut with a circular knife to the required size. As this machine is well known to persons conversant with this manufacture, I need not further describe it. And, lastly, I do further declare that I do not, under the said first-recited patent, claim as new the processes described by my said former specification, nor the mixing of colouring matters with the caoutchouc, nor the common solution of caoutchouc as a cement, nor the use of calenders, or of any other machinery, matters, or things that have been previously used or practised. — In witness whereof, &c.

THOMAS HANCOCK.

Enrolled July 23, 1838.

VULCANISING.

Specification of the Patent granted to THOMAS HAN-
COCK, *of Goswell Mews, Goswell Road, in the County
of Middlesex, Waterproof Cloth Manufacturer, for
Improvements in the Preparation or Manufacture of
Caoutchouc in Combination with other Substances,
which Preparation or Manufacture is suitable for
rendering Leather Cloth and other Fabrics Waterproof,
and to various other purposes for which Caoutchouc is
employed.* — Sealed November 21, 1843.

To all to whom these presents shall come, &c. &c. — The
nature of my improvement or improvements in the preparation or manufacture of caoutchouc in combination with
other substances, consists in diminishing or obviating their
clammy adhesiveness, and also in diminishing or entirely
preventing their tendency to stiffen and harden by cold,
and to become softened or decomposed by heat, grease, and
oil.

I will first describe the means by which I correct,
obviate, or lessen, the clammy adhesiveness of caoutchouc,
and caoutchouc in combination with other substances. And
I would first premise, that as the essential oils employed in
softening and dissolving caoutchouc are ultimately almost

entirely evaporated, I wish to be understood when speaking
of proportions, that the dry materials are meant, more or less
solvent may be used, according as it may be convenient
to employ the combination in a stiff and plastic state, or
diluted in any degree down to the consistence of painters'
varnish. I take ten pounds of caoutchouc, and pass it two
or three times between iron rollers, until a roughly uniform
sheet is obtained; I then take twenty pounds of silicate
of magnesia (sometimes called Venetian or French chalk
and talc), and reduce it to a fine powder, and I continue
rolling the sheet, shifting the silicate upon it as it passes
through or between the rollers, and I carry on this operation until the whole is well mixed in. I then work up
the mass into a state of uniform consistence by means of
the machine or apparatus commonly employed in making
what is called manufactured caoutchouc, which machine
is fully described in the specification of a patent granted to
me on the 18th day of April, 1837. If the mass is intended for cutting into sheets or other forms, I press it in
moulds, and cut it up in the manner commonly practised,
and well known to persons acquainted with such manufactures. When I require this combination in the form of
large sheets, and they are not required of very fine quality,
I pass it between rollers, beginning with them rather wide
apart, and gradually closing them each time of passing
through, until they produce the thickness I want. The
rollers may be used cold, or heated to about 80° Fahrenheit. When it is required to spread the combination on
cloth, either for the purpose of rendering a previous coating
of caoutchouc unadhesive or upon the surface of the cloth
itself, I proceed to soften the combination by the application
of a small quantity of solvent, so as to make it of the
consistence of dough or putty, in the manner set forth in
the before-mentioned specification of a patent granted to me
in 1837, in respect to caoutchouc ; and I spread it also in
the manner and by a spreading machine similar to the one

S

therein described. If very thin sheets or films are required, I spread the combination on cloth previously saturated with size, and proceed with this operation in the same manner as is set forth with respect to caoutchouc sheets, under a patent granted to me, bearing date the 23rd day of January, 1838. The combination in this state, if not of too soft a consistence, may be spread by means of iron rollers into sheets, either alone or upon cloth, so as to remain permanently attached to the same, or upon sized cloth, to be afterwards stripped off; the surface of the iron in contact with the sheet should be kept wet, to lessen or prevent adhesion. Instead of a smooth even face, I sometimes obtain impressions on the surface by pressure between plates previously engraved of the desired pattern, or by means of rollers, by engraving one or both, so as to produce the required graining, or device; these plates or rollers may be used cold, or if the consistence of the combination is hard and stiff, they may be heated to any required degree. If a dead or dull matted appearance is required, I pass woollen cloth or other suitable fabric through or between the rollers, in contact with the coated surface of the cloth or sheet, and afterwards strip off the fabric, which should be previously sized, as before stated. By these and similar means, I frequently give to sheets of the combination, or to fabrics coated with it, a very close resemblance to a variety of woven and other manufactures, some of which may be used as substitutes for the article imitated. For coarse and cheap articles a proportion of washed lime or fullers' earth, dried, and sifted very fine, may be mixed with the silicate, and the quantity of the latter proportionably reduced. I sometimes make the exterior of an article of a combination that will be inadhesive, and of fine materials, and the interior with a combination containing either more caoutchouc, or more of the lime or fullers' earth, as the case may require. The combinations hereinbefore described will be of a light drab colour, or of a darker shade, according to the proportion of the silicate; but if required of other colours, I mix with the silicate any suitable pigment commonly used for such purposes. If patterns are required to be printed on these manufactures, I mix the colour with a thin solution of caoutchouc, as mentioned in the specification of my patent of the 23rd day of January, 1838. When a dark colour is not objectionable, I employ asphalte instead of, or combined with, the silicate. The combination of asphalte with caoutchouc may be effected in the dry state by reducing the asphalte to a fine powder, and treating it in the same manner as before stated in respect to the silicate. Plumbago may also be introduced when the colour is dark. I find that from six to seven pounds of asphalte, according to its quality, to eight pounds of caoutchouc answers the purpose; if silicate also is used, the quantity of asphalte should be proportionably less. When this combination is required for spreading on cloth, or for thin sheets, I soften it, and treat it in the same manner as before described for the silicate; if it is wished, more asphalte or silicate may be added during this operation. Nearly the same results are obtained by dissolving the asphalte in coal naptha, and employing such solution instead of naphtha, or other solvent, in softening the caoutchouc, to bring it to a suitable consistence for the spreading machine; and if silicate is employed with the asphalte, it may be sifted in, as directed in respect of colouring matters, in the specification before referred to, of 1837. Raised patterns may also be produced on the surface by the means before described. These combinations may be manufactured into sheets and various other articles, or applied generally to caoutchouc and caoutchouc manufactures, where it is desirable to obviate the clammy adhesiveness of the surface; and I apply them for the purpose of rendering cloth more or less waterproof by means of the spreading machine before alluded to, and to other purposes, by means of a spatula, or

s 2

other convenient instrument, and in a more diluted state, in the manner of painting, or varnishing with a common brush. Of course the different proportions may be varied according to the purposes to which the combinations are to be applied. It may be necessary here to observe that all the substances employed in the before-mentioned combinations have a tendency to weaken, more or less, the elastic properties of caoutchouc, and in particular the proportion of silicate of magnesia may be increased until the caoutchouc is nearly deprived of those properties. Any person practically acquainted with caoutchouc manufactures will be able, from the descriptions already given, to obtain the proper consistence and quality, and to adapt the combinations to the various purposes for which they may be required.

It is well known that caoutchouc stiffens and becomes hard when exposed to a cold temperature; and that it is liable to become soft and decomposed by heat and exposure to the atmosphere; and by contact with oil or grease; that it is easily acted upon by solvents, and its elastic properties weakened by the means usually employed for its manufacture. I diminish or obviate these effects by intimately blending sulphur with caoutchouc during the process of its manufacture or preparation, and then treating the combination in the manner hereafter described. Sulphur may be blended with caoutchouc in various ways, but the following I find to answer the desired purpose: — I melt in an iron vessel a quantity of sulphur, at a temperature ranging from about 240° to 250° Fahrenheit, and immerse it in the caoutchouc, previously rolled into rough sheets, or cut to any convenient form or size, and allow it to remain until the sulphur has penetrated quite through the caoutchouc, which may be ascertained by cutting a portion of it asunder with a wet knife; if the operation is complete, the colour of the caoutchouc will be changed throughout to a yellowish tint: if there is only a margin of yellow around the cut part the operation must be continued longer, until the colour of the whole is changed; the sulphur adhering to the surface being scraped off, the caoutchouc will then have taken up a quantity of sulphur, from one-sixth to one-tenth of its weight. With caoutchouc thus prepared, I proceed with the further manufacture of it into the consistence of dough or putty, or making solutions of it, and spread it on cloth or into sheets in the manner already stated or referred to. Sulphur may also be blended with caoutchouc by reducing the former to a fine powder, and mixing it mechanically in the manner and by the means before mentioned for silicate of magnesia. Sulphur may also be blended with the surface of some articles, such as sheets of caoutchouc, by heating the latter to about 200°, and sifting and rubbing flour of sulphur on it. Instead of the preceding, I sometimes blend sulphur with caoutchouc by means of a solvent. In that case I saturate the solvent I mean to employ with as much sulphur as it will take up by boiling, and employ as much of this saturated solvent as will, after evaporation, leave the requisite proportion of sulphur before indicated blended with the caoutchouc. When this saturated solvent is allowed to cool, any excess of sulphur will fall down in crystals; if, therefore, for any purpose, it is wished to employ a larger proportion of sulphur, it must be kept hot. I prefer in this case to use oil of turpentine as the solvent. Either of the foregoing, or any other convenient mode, may be adopted for blending the sulphur and caoutchouc together, care being taken to ensure as much as possible a uniform mixture. The silicate of magnesia and the other substances mentioned in the first part of this specification may, if wished, be mixed with the combination of sulphur and caoutchouc in such proportions as may be necessary to obviate or correct the clammy adhesiveness before mentioned; but I wish it to be clearly understood that the combination of the silicate of magnesia with caoutchouc has the effect in all cases of lessening its elastic properties in proportion to the quantity of silicate employed. This combination of sulphur and caoutchouc, and of sulphur and caoutchouc mixed with the silicate of magnesia and the other substances before mentioned, may be applied to various purposes in the manner hereinbefore stated or referred to, and introduced by similar or other convenient means into caoutchouc manufactures.

Having described the methods by which I blend sulphur with caoutchouc, and the manner in which I apply the same to various purposes, I would here observe that the combination is still as soluble as before, and has not yet undergone the change or modification by which the improvement or improvements contemplated in this portion of my invention are carried out. When caoutchouc alone is to be operated upon, I find that the desired effect, which for brevity's sake I will hereafter call the change, may be produced by immersing the caoutchouc in melted sulphur, as hereinbefore mentioned ; I then raise the temperature to 300°, or from 300° to 370°, and continuing it so immersed for a longer or a shorter period, according to the thickness or bulk of the caoutchouc or the extent to which the change is to be carried; for instance, if sheet caoutchouc one-sixteenth of an inch thick is continued in sulphur at 350° to 370° from ten to fifteen minutes, the change before alluded to is produced; or, instead of so high a temperature, the sulphur is raised only from 310° to 320°, and the caoutchouc immersed in it from fifty to sixty minutes, the result will be much the same ; and if continued for two hours at the same temperature, the effect will be proportionably increased: and if continued longer, the caoutchouc becomes of a darker colour, and nearly loses its property of stretching ; and if carried still farther, turns nearly black, and has something the appearance of horn, and may be pared with a knife similarly to that substance. By the effect of this high temperature such a change in, or modification of, the properties of the caout-

chouc and most of its combinations will be produced, that
the elastic force or property of manufactured caoutchouc
to recover its form after being extended is greatly increased, and it will, after being so treated, resist to a considerable extent the action of heat, oil, and grease, as well
as the effect of cold, and be more capable of resisting the
menstrua by which caoutchouc is commonly dissolved.
I would here observe that the temperature and the time
which have been stated produce a degree of change suitable for many purposes, but may be varied discretionally.
And I would here make this general remark, that the higher
the temperature the shorter is the time required; and, on
the contrary, a lower temperature requires a longer time.
And I would here observe that the proportion of sulphur
is, to a certain extent, increased by contiued immersion;
and the time and temperature just indicated is equally
applicable to the combinations of caoutchouc and sulphur
with the silicate of magnesia and other substances mentioned in the first part of this my specification. The sulphur which adheres to the surface may be easily removed
by friction or scraping. I employ this mode of operating
in preference generally where practicable, particularly for
articles made of manufactured caoutchouc, such as sheets
and thread for elastic manufactures, whether made from
manufactured caoutchouc or of the raw material as imported. When the combination is spread upon cloth, or
attached to other substances capable of enduring the necessary temperature, I pass such articles over plates or
cylinders heated sufficiently to effect the change; if the
side on which the combination of caoutchouc and sulphur
is spread is brought into contact with the heated surface of
the plate or cylinder, the temperature and the time in
contact formerly indicated in respect of melted sulphur
will answer the purpose; and this is the mode I prefer in
cases where substances, such as leather, coated with the
combination, will not so well endure a high temperature;

but when the combination is spread on cloth or other substance which is to be brought into contact with the heated surface, the temperature will require to be raised, or the article longer exposed to it, according to the extent to which the heat is impeded by the thickness or quality of the intervening substance, which may be easily ascertained by testing small pieces as a guide.   In some cases, such as for the strapping on the seams of garments, it may be useful to employ hot irons to effect the change.   Another mode of submitting the articles to the necessary temperature, is by means of a stove heated to the required degree, which degree, and the proper period for the article to remain in it, must, as before stated, depend on the nature of the case : if it is sheet caoutchouc, or thread, or any similar article, the time and temperature before indicated will answer, according to the extent to which the change is to be carried, or the size or bulk of the articles to be operated on. If the article is partly composed of cloth coated on one side, as, for instance, a single texture of calico, rendered waterproof by a very thin coating of the combination, a temperature of from 290° to 300° will be sufficient, and the period for remaining in or passing through the stove from one and a half to two minutes.   If a thicker coat of the combination is laid on the calico, a somewhat longer time must be given ; and so in respect of a union of two plies of cloth, the time and temparature must be regulated according to the thickness of the textures and the interposed coating of the combination.   When a greater number of plies of cloth are required to be united, as in the manufacture of artificial leather for straps, card-backs, hose-pipe, and the like, I proceed as in the case just mentioned ; but if, from the thickness of the cloth, or the number of plies required, I conceive the heat will not readily or sufficiently penetrate the mass, I unite two folds of the cloth first, and pass them through the stove until the change is effected ; I then unite another fold to the former, and pass it again through the stove, and so proceed until the required number of folds are united; and, if required, I lastly coat the surfaces in the above manner, and finish by again submitting the whole to the stove. Another mode of obtaining the temperature for effecting the change is by immersing the articles in water or steam, under pressure, raised to the required temperature. I find that a very small quantity of boiled linseed oil, stearine, or spermaceti, introduced with the sulphur, communicates an agreeable smoothness to the surface. I would here also remark, that when surfaces are to be united, the union should be effected before the change takes place. When the combination is required for purposes where it could not be conveniently used of the consistence of dough, such as for saturating cloth, felt, or other similar purposes, or for coating uneven surfaces, I dilute it with solvent to any required consistence, and apply it with a brush, or other convenient means, and afterwards submit the articles to the influence of heat, in the manner already described, or by any other that may be convenient. The sulphur may, if required, be more or less discharged from the caoutchouc, after it has undergone the change, by submitting it to the known solvents of sulphur, of which I prefer a solution of the sulphite of soda in water, kept to a temperature of about 200°. Other and similar modes may be devised for carrying these improvements into effect. But enough has now been said to enable any person of common skill, and practically acquainted with caoutchouc manufactures, to follow out these operations with success. What I claim as my invention and discovery is,

Firstly, The combination of caoutchouc with silicate of magnesia, whereby manufactured caoutchouc is rendered free from that clammy and adhesive character which it usually possesses.

Secondly, I claim the modes herein described of combining asphalte with caoutchouc; and,

Thirdly, I claim the treating of caoutchouc (either alone or in combination with other substances) with sulphur when acted on by heat, and thus changing the character of caoutchouc as herein described.—In witness, &c.

THOMAS HANCOCK.

Enrolled May 21. 1844.

FOR OBTAINING FORMS BY MOULDS AND VULCANISING.

Specification of Patent granted to THOMAS HANCOCK, *Stoke Newington, Middlesex, Esquire, for Improvements in the Manufacturing and Treating of Articles made of Caoutchouc, either alone or in Combination with other Substances, and in the Means used or employed in their Manufacture.* Sealed March 18. 1846.

To all to whom these presents shall come, &c. &c. — My improvements consist in manufacturing articles made of caoutchouc, either alone or in combination with other substances, also in forming and giving to such articles specific and permanent shapes or forms; also in producing a perforated, perflable, or vesicular manufacture, and also in certain compounds of caoutchouc with other substances.

I would first premise, that in all the compounds and most of the articles comprised in this specification I employ sulphur and heat, as described in the specification of my patent of November 21. 1843, which process is now commonly designated " Vulcanising," and I shall, for brevity's sake, use that term in this specification, whether it is applied to the whole process described in my said specification, or only to the completion thereof by heat when sulphur has already been introduced into the compounds ; and by the term compound, I wish to be understood always to mean any of the combinations comprehended in this specification that may be severally most suitable for each purpose.

When I manufacture these compounds into articles requiring to be of a permanent shape or form, I make such articles in or upon forms, moulds, plates, or engraved surfaces or patterns, by pressing, fitting, placing, or moulding such compounds, previously prepared in sheets or otherwise, in or upon such moulds or forms, and allowing the articles to remain there whilst exposed to the vulcanising process, which effectually sets them permanently to the respective forms. In order to prevent adhesion to the mould, I employ silicate of magnesia, either by dusting it on in the form of a powder, or with a brush when mixed with water, applied either to the mould or the compound, as may be most desirable. In some cases I find it convenient to remove the articles from the moulds before vulcanising, in which case I submit them for a short time before removal to a temperature of from 220° to 300°, according to the size or bulk of the article, for which purpose I prefer a water or steam bath, under sufficient pressure to produce the required heat. When cold, I remove them from the moulds, and afterwards vulcanise them to make their forms permanent. The same process of obtaining ornamental surfaces by moulds may be resorted to with advantage when caoutchouc, without the process of vulcanisation, is employed. For some articles that are not of much bulk, I take sheets composed of caoutchouc only, and obtain impressions from moulds, in manner just mentioned, in respect of compound sheets which I remove from the moulds, and then apply a thin solution of caoutchouc, containing a large proportion of sulphur, to the surfaces of such articles, or I rub dry powdered sulphur upon them, or immerse them in a sulphur bath and then vulcanise them, or I saturate any suitable solvent of caoutchouc with sulphur, and apply successive coatings to these articles, and when dry vulcanise

them.   The same proceeding may be followed in making
similar articles in moulds from solutions of caoutchouc only.
If the solvent employed is previously sufficiently saturated
with sulphur at a temperature of from 300° to 320°, these
articles will be thereby sufficiently vulcanised for some
purposes without carrying the process farther, but I prefer
applying heat afterwards.   I sometimes employ a solution
of caoutchouc, or caoutchouc softened to the consistence of
dough by the addition of a small quantity of solvent, as
described in the specification of my patent of April, 1837
(blending or mixing sulphur therewith).   I pour the solution or press the dough into moulds, forms, or patterns, and
allow them to dry and then vulcanise them.   The thinner
solutions answer well for hollow moulds, into which I pour
a sufficient quantity to cover all the parts.   I then allow the
solution to drain out, and place the mould in a stove or
warm room; and when this first coat is dry I repeat the
operation until the desired thickness is obtained, and, when
perfectly dry, I vulcanise them.   If necessary, the moulds
may be made in parts, as is well known and understood.
In some cases I make the figures in or on parts of moulds,
and then cement the parts together, and replace the entire
figure in or on the moulds and vulcanise it.   In some cases
I make the first coating with the solution, and thicken up
parts or the whole with dough, prepared as before mentioned.   Instead of the foregoing, I sometimes make these
figures of the compound prepared in sheets of any desired
thickness, and join them up to the form of the mould.
When for an internal figure, the opening being secured, I
apply internal pressure, by admitting through a stop-cock,
steam, or air, or by any other convenient mode, so as to
obtain the requisite pressure to bring up the pattern, and
continue such pressure during the process of vulcanising.
Instead of steam, or air, I sometimes fill the interior of the
hollow figure with mercury, or any metal that will fuse at
the temperatnre employed in vulcanising the article; as the

pressure in these cases is obtained simply by the weight of the metal, the sheet of compound forming the hollow figure should be proportionably thin. It is generally necessary to make small holes in the mould for the escape of air from between the mould and the article or figure. In cases where great pressure can be applied, I find that in shaping the figure before it is put into the mould an approximation only is required to the form of the mould, the heat and pressure being sufficient to bring up the finer parts. Bottles, and other vessels of capacity, I make in the same way. These hollow moulds require to be made strong in proportion to the size of the work and the required internal pressure. Casts from flat moulds, such as bas-reliefs and engravings, I also sometimes make of sheet compound, and force it into the moulds by heat and pressure, and then vulcanise them. This mode, amongst other uses, is particularly applicable to produce printing surfaces; I press the material into the moulds with screw-cramps, or by presses, or other well-known means; these or any other convenient modes may be adopted. I mention these as examples sufficient to show the workman what is my design, and how it may be carried out. I raise figures on or emboss cylindrical and other similar forms or articles, by engraving or otherwise, producing raised or sunken designs on the external surface of cylindrical moulds. I then form tubes of the sheet compound of such a size as will draw closely on the mould. I then wind very tightly over the whole a strip of cotton or linen cloth, so as to produce sufficient pressure to obtain the impression, or the requisite pressure may be applied by any other convenient means, and I then vulcanise them; ,and when the article is withdrawn from the mould, I turn it inside out, which brings the pattern to the outside. I find that impressions perfect enough for some purposes may be obtained on sheets of the compound, and sufficiently set for vulcanising by means of engraved rollers heated to about 240°. The pressure must be as light as the nature of the article will allow, and the motion of the rollers slow, and I prefer them of large diameter. When I wish to prevent extensibility in any of these articles, I either mix fibrous matters with the compound, or I attach, by means of the sulphured solution before mentioned, previously to vulcanising, any suitable fabric, preferring cotton or linen. For some of the purposes to which the compound is applied, it is desirable that articles formed of it should be so manufactured as to admit of air or perspiration passing through it. To effect this in sheets or articles made of the compound that are not very thick, I take out, by means of hollow punches of any required form, as much of the substance as the case requires; and I do this either before or after vulcanising, but I prefer the former. Instead of punching, I sometimes make incisions in the articles, and keep these incisions a little on the strain, or open in the required direction during the process of vulcanising, or I otherwise puncture the sheet or article and follow the same treatment. The punching, incising, or puncturing, may be done by hand or by machinery, the chief care being required in preventing any closing of the openings during the vulcanising process; if the articles are ornamented by embossing, and it is desirable to conceal or obviate the appearance of these perforations, I form my design for such ornament so as that the openings may form part of such design. If it is wished to make any part of an article weaker at certain points, I find perforating or punching out the material a convenient mode of effecting this object to any required degree; the parts so weakened will be less liable to tear out if the perforations are made previously to vulcanising.

If for any purpose it is required that the sheets or articles should be punctured to a certain depth, but not through, I take two sheets, one punctured and the other plain, and unite them face to face before I vulcanise them. If for any purpose articles require either to be sewn together, or to any other material, I make the compound thicker in the part where the needle is to pass; or, previously to vulcanising, I place or insert a strip of cotton or linen coated with solution upon or between them.

For low-priced articles, I sometimes combine caoutchouc and sulphur with vegetable or Stockholm pitch; and when spread into sheets, or made up into other forms, I vulcanise the compound. The proportions may be varied very considerably, as well as the temperatures at which they are vulcanised, but I find the following to answer well: eight parts caoutchouc, two sulphur, three pitch; or eight parts caoutchouc, two sulphur, one pitch, submitted to a temperature of 290° for an hour, to prevent blistering and porosity if necessary, I employ pressure, by means of screw-cramps and plates, or otherwise, during the vulcanising. This material is applicable to railway packing, and other rough uses. I also combine and vulcanise in the same manner caoutchouc, sulphur, and resins, preferring, on account of its cheapness, the common resin of commerce. The proportions and temperature, as in the case of pitch, may be varied, but the following I find to be useful for many purposes: sixteen parts caoutchouc, two parts sulphur, six parts resin; or sixteen caoutchouc, four sulphur, two resin. These compounds may be submitted to the same treatment as in the case of pitch, and are applicable to similar uses.

For some purposes I also combine caoutchouc and sulphur with wood or cork dust, or fibrous substances, such as hemp or flax, and any other suitable material, cut into short lengths, and vulcanise such compounds, either in blocks, or spread, or otherwise wrought into sheets, or formed into figures, or embossed or ornamented, as before described.

When it is desirable to obtain from caoutchouc a great amount of elasticity, to counteract the collision of great forces, or heavy bodies, as in the case of carriage-springs, and the buffers of railway-carriages, and for other similar purposes, I form the material in such a manner as to obtain a large amount of divisions or openings, so as to expose and bring into action a great extent of moving or movable surface; and I do this either by cutting or otherwise shaping any given quantity of material into parallel or suitable forms, either square, hexagonal, or octagonal, or into solid or open hollow cylinders, corrugated sheets, or any other suitable forms; and before vulcanising, build them up and cement them together, either across each other at right angles, or diagonally or otherwise, always leaving as much space between each as to admit of the free action of the surfaces when compressed. Hollow cylinders, or other hollow forms, I sometimes cement together, laying them longitudinally side by side, and in either case continue to build up, until I obtain the dimensions necessary to produce the required amount of elasticity. Similar structural forms may be advantageously applied to other uses. When it is desirable that extensibility in any direction should be prevented, I apply, by cementing on or between some of the layers, when building up the structure, linen cloth of suitable strength, coated with solution, as before mentioned.

I have in a former part of this specification described a mode of manufacturing figures in strong hollow moulds; I pursue the same methods in forming chambers to contain air for resisting pressure and blows, when used for carriage-springs, buffers, and other elastic surfaces for railways, and when stuffing or padding beds, seats, cushions, and other surfaces requiring elasticity.

In some cases, such as railway buffers, it may be necessary to make provision against accidental ruptures in these chambers, which may be done by enclosing several one within another.

I effect this by making them in parts, with flanges at the joints; and after vulcanising, riveting, or bolting the parts together, or in any other convenient way, the sizes

T

of such chambers to be so regulated as to leave sufficient space for the projection of the flanges: when necessary, I make apertures for the purpose of inflating these chambers to any required degree, in the manner well known in similar manufactures. A series of cylindrical or other forms for containing air may be cemented together, and then vulcanised, the openings by which they are to be inflated being left open during the process of vulcanisation.

For mattrasses, beds, life-preservers, cushions, pads, carriage-linings, parts of saddles, and horse-collars, and other the like purposes, I apply these air-chambers, of cylindrical or other forms, by enclosing them in cloth or leather cases, divided into compartments to receive them, or in cases of the vulcanised material, either plain or ornamented. I have mentioned in a former part of this specification that I combine pitch, resin, and various other substances with caoutchouc, previously to vulcanising; and I wish now to state that I do this by the means described in my former specifications, to which I would add, that in combining pitch and resin with caoutchouc by mechanical power, I find it useful to introduce water into the machine during the operation.

The moulds I make of glass, when it can be conveniently applied, tin, type-metal, porcelain, and highly-vulcanised caoutchouc. They may be made of any other suitable material, but I prefer the above. If for smooth or polished surfaces, I use sheets of plate glass; if for ornamental purposes, I cast them of glass of the required pattern or form.

I have now stated the manner in which I manufacture articles of caoutchouc, either alone or in combination with other substances; the means by which I give permanency of shape or form to such articles, and the compound substances employed in the various manufactures. I have mentioned also, in several instances, the purposes to which the compounds are suited, and modes of application; yet I have done so only by way of example, the modes and compounds being well adapted to a great variety of other articles and purposes, such as embossed fancy articles of dress, bracelets, collars, ornamental edgings and borders, military epaulettes, belts, and other similar articles. Imitations of crape, and of cord lace and fringe, for coach-linings and other purposes, picture and other frames, ornaments for the decoration of cabinet furniture, and hangings and draperies, either elastic or attached to cloth.

What I claim as my invention is,

Firstly, the making, forming, or shaping articles from the combination of caoutchouc with other substances, as hereinbefore described, in or upon moulds, plates, or forms, and retaining such articles in or upon such moulds, plates, or forms, during the process described in my specification of 1843, and now commonly called vulcanising, whereby the form of such articles is rendered permanent.

Secondly, the making, forming, or shaping articles of caoutchouc in or upon engraved or otherwise ornamented plates or moulds; and after forcing the caoutchouc into such moulds by pressure and heat, submitting the whole, by means of a water or steam bath, or any other suitable mode, to a high temperature, whereby the articles are sufficiently set to be removable from the moulds, and which may be afterwards, if desired, subjected to the vulcanising process.

Thirdly, the manufacturing articles, by combining caoutchouc with vegetable pitch, resin, wood, and cork dust, and fibrous substances, and subjecting them to the process of vulcanisation. — In witness, &c.

THOMAS HANCOCK.

Enrolled September 18. 1846.

CONVERTING APPLICATIONS.

Specification of the Patent granted to WILLIAM BROCKEDON, *of Devonshire Street, Queen Square, Gentleman, and* THOMAS HANCOCK, *of Stoke Newington, Gentleman, for Improvements in the Manufacture of Articles where India Rubber or Gutta Percha is used.*—Sealed November 19. 1846.

To all to whom these presents shall come, &c. &c. — The improvements we have made in the manufacture of articles where India rubber or gutta percha is used, consist of peculiar means of applying these substances to a variety of purposes to which they have not heretofore been so applied, by means of the processes described in the specification of a patent granted to Mr. Alexander Parkes, dated the 25th of March, 1846, entitled, "Improvements in the preparation of certain vegetable and animal substances, and in certain combinations of the same substances, alone or with other matters." The processes enumerated in this patent produce certain changes in the qualities of caoutchouc and gutta percha, some of them similar to those produced by sulphur and heat in the process now termed "vulcanising," in others purifying and colouring those substances, and by these means rendering them suitable to a great variety of purposes.

In this specification we propose to follow Mr. Parkes generally in calling these substances by the names of caoutchouc and gutta percha, which will be found convenient; but we wish to be understood, that when using these terms we intend to comprehend all those peculiar hydro-carbon substances known to botanists as a vegetable constituent under the various names by which caoutchouc or India rubber is known. Some of these are derived from the country from which they are obtained, as Para, Assam, West Indian, Madagascar, Java, &c.; some are names given by the natives, such as saikwah, jintarvan, gutta tuban, gutta percha, doll, &c.; others from the condition in which it is received, as liquid, cake, bottle, root, sheet, scrap, &c.; and they differ also in colour, some being black, others white, red, brown, yellow, mottled, &c.; many of these varieties are reported by Dr. Roxburgh, Lieutenant Veith, and others, in the " Transactions of the Agricultural and Horticultural Society of India ; " and these products also vary in their hardness, from that of the solidity of wood to that of the soft and viscous state of birdlime, which does not harden naturally. And we would state that India rubber, or the peculiar property of the vegetable matter first introduced into this country under that name, consists in this, that it is tapped from a tree or plant, and for the most part it coagulates ; part of the fluidity being evaporated, the product thus obtained is not soluble in water, and in this particular it differs from ordinary gums, sugar, and starch ; and, further, India rubber and all other of the vegetable products having the properties above mentioned will, on distillation, produce caoutchoucine. These substances, under whatever name and however mixed and compounded, are all liquefied by the same solvents, and all require, in their preliminary manipulations and manufacture, the same or simple modifications of the same modes of treatment as ordinary India rubber, these being solvents ; also destructive distillation, to obtain the spirit called caoutchoucine ; also to rollers, masticating, spreading, cutting, and other machines ; also to processes for colouring, em-

embossing, printing, moulding, &c. &c., which are well known, and been before described in the specifications of other patents; amongst others, in the specifications of the patents granted to the within-named Thomas Hancock, dated the 18th April, 1837, 23rd January, 1838, 21st November, 1843, and the 18th March, 1846, as well as to the first-named patent of Mr. Parkes. The details of manipulation, which are described in these patents, will be found amply sufficient for the guidance of any workman conversant with such manufactures, and we have found the proportions enumerated therein to answer well for general operations and for the several purposes for which they may be required. With regard to dissolving varieties apparently different, such as ordinary India rubber and gutta percha, it may be necessary to mention, that the process is precisely the same in all respects with both during summer; and although the former may be dissolved at any of the ordinary temperatures of the atmosphere, yet the process is facilitated by heat, and may, therefore, at all times be carried on advantageously in the same room with gutta percha, which should have a temperature of from 80° to 90° of Fahrenheit. It is also necessary to keep both the spreader and the bed of the spreading machine heated when using gutta percha or compound, and the same remark applies to the masticating machine, say from 190° to 200° Fahrenheit. The principal defect of gutta percha consists in this, that although much harder than ordinary India rubber (meaning by that name that description which was first brought into this country), when at low temperatures, it becomes inconveniently yielding and plastic at comparatively low temperatures when compared with ordinary India rubber; but it is found, by treating it according to Mr. Parkes's patent, this property is materially obviated, both in respect to its hardness and capability of bearing heat.

We would here observe, that when we hereafter adopt

from Mr. Parkes's specification the word " change," we use
it to denote the same process or processes, and also that by
the word " immersion," the mode of producing the change
by immersing articles in solvents capable of producing such
" change " is meant, which process we generally prefer.
We render leather, cloth, linen, silk, and other fabrics and
materials partially or entirely waterproof by coating their
surfaces or uniting two or more of them together with
caoutchouc, gutta percha, or a compound of these matters
in a state of solution or otherwise, as described in the
patents of the said Thomas Hancock before referred to ; and
we make the coated surfaces of these cloths either plain,
coloured, embossed, printed, or otherwise ornamented, and
then produce the "change" by " immersion; " these manufactures differing from those of Mr. Parkes and Mr. Hancock's inventions only inasmuch as operating on the manufactured article in place of acting on the raw material of
caoutchouc, gutta percha, or the compounds of those
materials.   A convenient mode of immersion with printed
or dyed fabrics coated on one side only is to join up the
selvages the whole length and the ends, and render the seam
waterproof, and then immerse it in this bag-like form. When
it is necessary to protect fibrous and other substances liable
to injury by contact with the changing solvents, we coat or
saturate them with glue-size, and remove it afterwards by
rinsing in warm water, or we employ an aqueous solution
of lac, which we remove afterwards by any suitable alkaline
solution.   By the same means we stop out the effect of
the changing solvent in any part of an article formed of
caoutchouc, gutta percha, or a compound thereof.   These
manufactures we introduce into a great variety of articles,
such as cloaks, capes, overalls, fishing stockings, collars,
stocks, hats, caps, bonnets, hat linings, hat bands, aprons,
and other articles of dress, or to be worn about the person ;
also, table covers, wrappers, carriage roofs, seats, and linings,
portable baths, diving dresses, life preservers, beds, cushions,

T 4

pads, and other pneumatic articles, printers' blankets, sieve cloths, card backs, draperies, hangings, covering walls.

These articles are made up by modes similar to these commonly practised in making up caoutchouc goods. When these articles require seams, or to be otherwise sewed together, the waterproofing substances (solutions of the above materials) employed to such parts will require afterwards to undergo the "change" by applying the converting solvent with a brush or otherwise. We sometimes make up garments or other articles of dress, such as gloves, gaiters, shoes, boots, leggings, galoshes, overalls, aprons, portmanteaus, and other similar articles, either of leather, cloth, or other material, and then apply coatings of caoutchouc, gutta percha, or of their compounds, in a state of solution, coloured or plain, by dipping or by hand, with a brush, or other means, and afterwards immerse them to obtain the change.

We would here remark, that although gutta percha is improved by the "change," in respect to its elasticity, it is still inferior in that respect to ordinary India rubber, and still possesses so little elasticity that it should be introduced sparingly, if at all, when that quality is required.

Articles intended for inflation, such as beds, cushions, pads, and other similar manufactures, we first make in the manner commonly practised in similar caoutchouc manufactures, employing either that substance or gutta percha, or a compound thereof, and "immerse" them afterwards. We prefer caoutchouc to gutta percha, the latter being too rigid for most of these purposes. When the exterior of the article is required to be of cloth or other texture, we protect the fabric when producing the "change" from the action of the changing agent, by forming the air-proof lining in such a manner as to cover the whole interior surface of such fabric, and obtain the "change" by pouring in the changing solvent, allowing it to remain only for the necessary period. When the exterior of the article is to be formed of caoutchouc, gutta percha, or a compound thereof, we stop up the orifice and immerse the article. And in all cases the colouring, embossing, printing, or otherwise ornamenting, we prefer to execute previously to effecting the " change."

We also manufacture vessels intended to contain air, water, or other fluids, composed entirely of caoutchouc or gutta percha, or of a compound thereof, by the modes described in the patents before referred to of the within-mentioned Thomas Hancock, and afterwards we immerse them to produce the change.

We also manufacture caoutchouc, gutta percha, or a compound thereof, with or without gritty or colouring matters and fibrous substances, and form them into sheets of any required thickness, by means similar to those described in the patents of the said Thomas Hancock before referred to, and employ them in the formation of any of the articles herein described, producing the " change," either when in the form of sheets, or after making them up into such articles as may be thought most convenient. From these sheets, whether combined with fabrics and fibrous and other substances or not, we manufacture straps for driving machinery ; deckle straps, reins, traces, and other parts of harness, horse collars, horse-shoe linings, horse furniture, such as knee-caps, fetlock boots, parts of saddles, and saddlery ; soles of shoes and boots, portmanteaus, balls, belts, gaiters, trouser straps, braces, shoulder straps for stays, waistcoat and waistband springs, shoe and boot fastenings, shoes, boots, galoshes, uppers, quarters, and vamps; air-chambers, bottles, and other vessels for containing fluids ; printers' furnishers, covering and lapping rollers, bowls, and other similar articles ; roofing, sheathing, washers for water, steam, and other joints; hose pipe, and tubing ; railway valves ; and packing block or springs, to prevent the recoil of guns, pump valves, and buckets ; covering stoppers and bungs ; covers of pickle jars ; capsules for bottles ; bandages, knee-caps, ligatures, and other surgical

apparatus ; a variety of embossed articles, such as fancy
articles of dress, bracelets, ornamental edgings and borders,
imitations of crape, fringe, and lace ; picture frames, ornaments for the decoration of furniture, forms and impressions to print from type.

We also manufacture cushions for billiard tables, by
uniting any number of sheets together, either entirely of
caoutchouc or of a compound of caoutchouc, gutta percha,
or intermixtures of sheets of both in alternate layers, to
modify the degree of elasticity, and immerse them to produce the change.   These sheets we also apply to cover and
protect plates of metal, and to the lining of metallic and
other vessels, and to chests and tanks of wood or other
material, effecting their union either by heat or by means of
the ordinary solution of caoutchouc, or of the compounds,
and produce the change after the sheets are applied, or if
the " change " be first produced to the sheets, we apply
the cement hereafter described.

From these sheets we also manufacture springs for
carriages and railway buffers, in the forms and by similar
means to those described in the patent of the said Thomas
Hancock of 1846, uniting the parts by means therein
directed ; or, as regards gutta percha and the compound
thereof, by the means hereafter mentioned ; and we then
" immerse" them to produce the change.

We also manufacture caoutchouc, gutta percha, and
compounds thereof with thread, and produce new manufactures of such threads by combining the changing process
of Mr. Parkes's patent in this manufacture; and we produce
different degrees of elasticity by varying the proportions of
the compound.   The time of applying the changing process
may be either before or after cutting the substance used
into thread.   We prefer, however, to make sheets of the
desired thickness of the thread, then to obtain the
" change," and then to cut the same into thread ; which we
do by coiling a sheet thereof around a cylinder of wood, or

other fit material, using a solution of shellac over the whole surface of the sheet, by which the coiled mass will be retained together. This cylinder we put on centres and cause it to revolve against a knife constantly supplied with water, by which successive discs of thread are cut off, the cement being afterwards discharged by boiling in a solution of potash. When the sheets are of considerable thickness for coarse thread, we find it only necessary to use the cement towards the outer coil or end of the sheet; India rubber or gutta percha thread, or thread of their compounds, may be made into cords, ropes, braidings, plaitings, webs, whips, and other similar articles, and then immersed ; by which we not only effect the change, but firmly unite all threads into one mass, and in doing so different coloured threads may be combined. The handles of whips or parts of the articles may have wood or other material introduced during their manufacture, to give those parts additional strength or stiffness. If elastic thread is first woven up with other thread of animal fibre, we take the woven fabric in the elastic state and extend its length, and retain it stretched during the immersion, and also after it is removed, until the solvent taken up is evaporated ; the thread when liberated will contract considerably.

We manufacture caoutchouc, gutta percha, and the compounds thereof, into various forms and patterns, by making such forms in or upon moulds, plates, and engraved or otherwise wrought and figured surfaces, sunk or in relief, by means similar to those described for caoutchouc in the aforesaid specification of Thomas Hancock, and afterwards immerse them, either before or after removing them from the moulds or forms. When immersing moulded or other articles which have fine impressions, or are of a delicate or light construction, we find it convenient to give them a dip in the changing solvent, and out again immediately, to harden the surface, and when dry to immerse them for the required period.

We manufacture gutta percha or compounds thereof into gun or pistol-stocks, umbrella, knife, sword, and other handles, by means of moulds, on which we engrave any pattern or design; or we make them of any desired colour, or colour them after leaving the mould. We sometimes form a foundation of wood, or metal, or other material, which we introduce into the interior of such articles before it is moulded, and when all is completed we immerse them. If the article is made entirely of gutta percha, without the use of solvents, we proceed to operate in the manner directed in the notice issued by Dr. Montgomery, in November 1843, when he introduced this substance into this country through the Society of Arts, by whom several manufacturers and other persons were furnished with samples and an account of his mode of treating it. We cannot give it better than in his own words : —

" The gutta percha when dipped in water near boiling can then be readily united, and becomes quite plastic, so as to be formed, before it cools to 130° or 140°, in any required form, which it retains at any temperature below 110°. We take as much of it as is necessary, throw it into hot water, when it soon softens and becomes as plastic as putty, when it can be moulded as required."

It is desirable to warm the moulds to about a blood-heat. When simple gutta percha is required to be made into sheets, the process is of the most simple kind; as it is only necessary to bring it to this plastic state by heat, and treat it by the means directed in similar operations with compounds in the specification of the said Thomas Hancock's patents of 1837 and 1838, before referred to ; they may then be immersed, or they may be first employed for making up any article, and immersed to produce the " change :" thus, by combining the mode of making sheets invented by the said Thomas Hancock with the process of Mr. Parkes, an improved manufacture of sheets of gutta percha will result.

Black-lead, gritty, or colouring matters, and fibrous substances, may, if wished, be worked in previously. To prevent the gutta percha or its compounds adhering to the rollers, we cover them with calico or other fabric, and keep the fabric moistened with a solution of soap or soda. The sheets are more evenly rolled and buckle less when the gutta percha contains colouring matters, such as ochre or plaster of Paris. We sometimes use gutta percha not previously formed into sheets, and when softened by heat to the consistence of putty as recommended by Dr. Montgomery, and employ it in the formation of a great variety of articles by moulding with the hand or otherwise, and then immerse them to produce the " change." When gutta percha alone or compound thereof containing but a small proportion of caoutchouc is formed into blocks by the means before referred to, shavings and sheets can be readily taken therefrom with a common joiner's plane, and such planes may be made of any required size and strength, and set to produce varied thicknesses of sheets. We form gutta percha or compounds into cylindrical blocks, and cut, by the plane or other cutter, straps helically therefrom; these narrow sheets when immersed, and thereby " changed," form straps for driving machinery and other uses : but we would state, that the fault of gutta-percha driving-straps is their being liable to suffer by abrasion and heat; this, however, is much obviated by the " change," and, in some cases, these articles may with advantage receive a coating of caoutchouc or compound thereof previously to immersing them.

Blocks of this material cut up more freely when they contain a considerable portion of finely-pulverised pigment or earthy matter, such as ochre or pipe-clay.

These blocks may be drilled with a common drill; portions of them may be turned in a lathe; screws cut out, and holes tapped for them, by common screw tools. Gutta percha and compounds thereof may be carved and morticed, and various articles of furniture made from it, and then immersed to produce the " change."

We find that some articles formed of gutta percha are improved by being coated with caoutchouc, coloured or plain, and then immersed to produce the " change," and the same remark applies to the coating of caoutchouc with gutta percha or compound.

For any articles of great delicacy, and when a light colour is desirable, we purify the gutta percha in the manner described in Mr. Parkes's patent.

We employ caoutchouc, gutta percha, or compound, in binding books, portfolios, and similar articles, in the manner commonly practised with caoutchouc, and immerse the necessary parts or otherwise apply the changing solvent to them. By this means we obviate the great defect in caoutchouc bookbinding, that of stiffening in cold weather, as we render the backs always alike flexible and elastic. Leather or cloth coated and embossed, coloured, printed, or otherwise ornamented, as before described, we employ for the surface of the covers of books and other similar articles. We manufacture a material from caoutchouc, gutta percha, or compound, suitable for many purposes, by cutting out with punches or by other means, from sheets of different colours, patterns or designs, so formed and arranged that the pieces cut from one sheet of one colour, say red, shall exactly fit and coincide with that cut from another, say black, in the manner of buhl-work ; these pieces of different colours are placed one within another and cemented to cloth or other material, and submitted to pressure with a gentle heat, and then immersed to produce the " change." Instead of placing these pieces within each other, they may be placed singly upon plain or coloured sheets, or coated cloth, so as to form patterns sunk or in relief, and then immersed to produce the "change;" any of these when cemented to cloth are suitable for table-cloths, or to be glued or attached to furniture and other uses. When made thicker and from bolder designs and of cheaper colours, or cloth or other material, some of them are suitable for covering floors, staircases, &c.

We cover or completely case and envelope, so as if necessary to seal hermetically, articles or vessels made of wood, metal, leather, paper, cast, plaster, cord, string, and other substances, by dipping them in a thin solution of caoutchouc, gutta percha, or compound; and, when dry, immerse them to produce the " change." The coatings may be repeated before immersion until any required thickness is obtained. Other articles variously compounded, such as of treacle and glue, or the like matters, after being made of the required form, may be dipped into the solutions of caoutchouc, or of gutta percha, or of their compounds, and thereby rendered impermeable, and then immersed to produce the " change." Various substances, such as wood-shavings or dust, cork, leather, pulp, and similar matters, mixed and cemented with glue, paste, or gum, and formed into any desired shape, and, when dry, may be dipped in a solution of caoutchouc, gutta percha, or compound thereof, to produce the " change." We manufacture an article very much resembling sponge by mixing with a solution of caoutchouc, gutta percha, or a compound thereof, a solution of chloride of sulphur, as described by Mr. Parkes; after a short time the whole becomes coagulated or gelatinised, it is then to be exposed to a temperature of about 212°, in water or otherwise, until the solvents are evaporated, and if greater stiffness is desired it may be immersed in the changing solution. We prefer caoutchouc to gutta percha for these purposes. Another mode of proceeding is to subdivide into larger or smaller pieces either caoutchouc or a compound, or both, preferring the former, and filling rather loosely with these pieces a vessel of any open or net-like construction of the required form and immerse it to produce the " change," allowing the superfluous solvents to run off; by which means the pieces will be sufficiently united to form a compressible and elastic mass suitable for cushions, pads, and other purposes. We manufacture hose-pipes and tubing of caoutchouc, gutta percha, and compound in various ways. We take threads made of either of the above, of a size proportioned to the hose, and braid it upon a core formed of rope which has previously been coated with treacle and glue, or glue and whiting, and made perfectly smooth. The braiding may be repeated, or a coating of either of the solutions may, if necessary, be given, and when dry rolled under pressure with a gentle heat ; we finish by immersing the whole, and thereby produce the " change " and unite all the coatings ; the core is afterwards removed by boiling in water. For fancy tubing the threads may be of various colours. By another mode we take woollen or worsted yarn, of a size proportioned to the strength of the required hose or tubing, and saturate and coat it with a solution of caoutchouc, gutta percha, or a compound thereof, until the fibres are all covered, and when dry we braid it upon a core as above ; we then roll it under pressure with heat, or, if necessary, give it a previous coat or two of either of the solutions, and then immerse it to produce the " change." We manufacture these articles also by winding these threads or narrow strips spirally round the core, keeping the edges quite close, and, if necessary, wind another tape or thread over the first in the contrary direction ; we then roll them well under pressure and heat and immerse them to produce the " change," removing the core as before mentioned. Woven hose and hose and tubing manufactured of leather or felt, coated or lined with the above or any other mode in which caoutchouc or gutta percha are employed, we treat by immersion to produce the " change," either during their manufacture or when finished. We also coat the exterior or interior surface of ordinary caoutchouc hose and tubing with either of the above solutions in their ordinary state or coloured, and immerse them afterwards to produce the

" change."   Silk or wool, or other animal fibre, is most
suitable when used in combination with these substances
in articles intended for subsequent immersion to produce
" the change."

For some purposes we cover or coat the surfaces of
caoutchouc, gutta percha, or compound, with ground flock
or other suitable powdery substance by giving the article
to be flocked a coating of caoutchouc varnish, and then
dusting or spreading the flock or powder over it; when
dry we immerse the article to produce the " change." Such
surfaces, among many other uses, are particularly suitable to
the lining of vamps and the interior of shoes, galoshes, &c.

Sheets or other articles formed wholly or in part of vulcanised or ordinary caoutchouc, we colour by dipping, or
otherwise coating them with coloured caoutchouc varnish,
and then immerse them for a short period ; thus producing
coloured surfaces to vulcanised India rubber or caoutchouc.
As the union of these substances, caoutchouc and gutta
percha, can be so readily effected before the " change," we
prefer to do so when convenient; but when joinings are
necessary to be made after the " change," we employ the
cement by which we unite vulcanised caoutchouc, composed
of vulcanised caoutchouc melted by heat, and when nearly
cold, add to and mix with it an equal quantity of the
changing solvent.   These are to be well stirred together at
a gentle heat.   We prefer to apply it warm in thin coatings,
and when dry, if necessary, we give both surfaces a second
coat; the union should take place when the cement is nearly
dry, and the parts kept under gentle pressure for some
time in a warm place : it is difficult to describe the exact
condition required in the state of the cement as the best
moment of junction, but it should be nearly dry; a little
practice will, however, enable a workman to accomplish the
object.   When articles made of caoutchouc, gutta percha,
or a compound thereof, are required of considerable thickness, Mr. Parkes recommends a weaker solution of the

chloride of sulphur, and the article to be immersed for a somewhat longer time; and we have also found that caoutchouc and the compounds may for many purposes be made of any required thickness, by uniting sheets or other forms together, by pressure, as they come wet or only partially dried out of the immersion.

If the surface of any inflexible or other article, coated or otherwise treated with or made of caoutchouc, gutta percha, or a compound thereof, requires in whole or in part to be more or less indurated, this may be done by frequently immersing the article in the changing solvent (allowing it to become nearly dry each time), until it becomes, if necessary, as hard or harder than ivory, and may then be filed and wrought with tools, and highly polished for such purpose. We have found that the proportion of chloride of sulphur in the changing solvent may be increased so as to expedite the hardening.

For producing the "change," we wish always to be understood to prefer employing the chloride of sulphur, dissolved in bisulphuret of carbon or other fit solvent of caoutchouc, in the proportions mentioned by Mr. Parkes; but do not confine ourselves thereto, immersing the articles for the periods mentioned by him, varying them in both respects according to the thickness of the articles or the degree of change we wish to produce.

We have also found that equal parts of bisulphuret of carbon and coal naphtha answer well, but in this case the naphtha must be very pure.—In witness, &c.

WILLIAM BROCKEDON,
THOMAS HANCOCK.

Enrolled May 19, 1847.

PRINTING.

Specification of a Patent granted to THOMAS HANCOCK, *Stoke Newington, Middlesex, Esquire, for Improvements in Fabrics elasticated by Gutta Percha or any of the Varieties of Caoutchouc.* — Dated 2nd November, 1847.

GUTTA PERCHA, and all the varieties of caoutchouc and combinations of these varieties, are in their natural state more or less affected by differences of temperature, but means have been discovered by which such variations in their elasticity are overcome, and the improved substances are thereby better suited to the manufacture of fabrics elasticated by gutta percha or any of the varieties of caoutchouc.

Elastic, braided, woven, or other fabrics, whether of cotton, wool, silk, or other substances, have hitherto been generally manufactured either white or of some dyed or undyed ground, or if ornamented, the colours were introduced in the course of weaving, either in stripes by the common loom, or having patterns produced by looms of a more complicated character and at a proportionably increased cost, in which patterns were of necessity formal and often ill-suited to the articles to which they were applied.

My improvements in these manufactures consist in printing their surfaces, and thereby enabling the manufacturer to produce at much less cost ornaments of a superior quality and of every variety of taste or pattern, colour or design.

Sometimes elastic fabrics are formed by cementing to caoutchouc, when in an extended state, silk, cotton, leather, or other covering, the surfaces of which I ornament also by printing them.

Although caoutchouc which has been rendered permanently elastic is much more suitable for these manufactures than ordinary caoutchouc, yet articles, when elasticated by the latter, I also ornament by the same means.

These fabrics or articles may be printed by any of the well-known means employed in printing manufactured goods, such as blocks, cylinders, or any suitable apparatus, and with the colours and processes ordinarily used in printing the various materials of which the exterior of the combined fabric may be formed.

These fabrics or articles I generally print when in their contracted state; the effect of which is that the pattern is always the same, when after extension the article is allowed to recover its original dimensions.

I do not, however, always print these fabrics when in a state of contraction, but extend them, more or less, and print them when in the extended state; by which means I obtain from some designs a novel effect; the articles on being allowed to contract concentrating parts of the design or pattern and changing its figure.

There is another advantage in this mode of proceeding. — Elastic articles, such as garters, bracelets, bands, &c., are, when actually in use, generally somewhat extended; if, therefore, the pattern presents its most perfect phase, when of the dimensions or proportions originally designed, I stretch the article when printing it to about the same extent it would assume when in use, and thereby secure the desired result.—In witness, &c.

THOMAS HANCOCK.

Enrolled May 2, 1848.

VULCANISED SOLUTIONS.

Specification of a Patent granted to THOMAS HANCOCK *and* REUBEN PHILLIPS *for improvements in the Treating or Manufacture of Gutta Percha or any of the Varieties of Caoutchouc.* — Dated December 30, 1847.

OUR improvements consist in dissolving gutta percha or any of the varieties of caoutchouc, or reducing them to a soft, pulpy, or gelatinous state, after they have undergone the process of " vulcanisation " or " conversion; " and also in heating unvulcanised solutions, or preparations of these substances, so as to bring them to a vulcanised state. Also in improvements in moulds employed in the manufacture of articles from these substances.

The terms " vulcanising " and " converting," as applied to these matters, are now well known as designating certain improvements in the manufacture of caoutchouc, whereby it is rendered less liable to be affected by variations of temperature; and our improved solutions, when dried, partake, more or less, of the same property; the first-named process is described in the specification of a patent granted to the said Thomas Hancock, dated the 21st November, 1843; and the second, in the specification of a patent granted to Alexander Parkes, dated 25th March, 1846.

When using the term " gutta percha," or any of the varieties of caoutchouc, we wish to be understood as comprehending all those substances known by the Indians or natives in the country where it is produced under the names of saikwah, jintawan, gutta tuban, gutta percha, dollah, &c.; and in the markets of this country by similar names, and others, such as caoutchouc, liquid, cake, bottle; and for the sake of brevity we will, in this specification, designate the different varieties under the general term " caoutchouc," and our improved solutions by the term " vulcanised solutions." In operating on these substances, when in a vulcanised or converted state, we generally take the waste or cuttings of these materials and pass them between rollers, or otherwise reducing them to shreds or sheets, and boil them in oil of turpentine until they are dissolved, keeping the mass well stirred during the operation.

Other solvents may be used, such as coal naphtha and other of the essential oils; but the necessary temperature in such cases cannot well be attained without employing closed vessels: we, therefore, prefer using the oil of turpentine.

As these cuttings and waste vary greatly in quality, it is scarcely possible to state any precise rule as to the proportions of substance and solvent; nor is it at all material. A good mode of proceeding is to put into the vessel, in which the vulcanised solution is to be made, as much solvent as will cover the shreds of caoutchouc, and then add about one third, more or less, of the solvent; and when, in the course of dissolving, the solution becomes of the desired consistence, to pour it off, and then make up the next charge according to the remaining deposit, or an estimate of the results obtained. In general we find that, although a thick pulpy or gelatinised solution may be obtained by one operation, we prefer making the solution of moderate consistence, and, if necessary, evaporate away the superfluous solvent by any of the well-known means.

With regard to varieties in the qualities of the cuttings or waste, we can only furnish some general remarks for the guidance and discretion of the workman, who will soon acquire sufficient experience to enable him to carry on the operation with success. If the vulcanising or converting processes have been carried so far as to render the substances hard and horny, they may be nearly insoluble, or require so long a time as to render them scarcely worth operating upon; but if the vulcanising or converting has been carried so far only as to render the substance strongly elastic, or if it retain a degree of flexibility not very different from its original state, it will dissolve at the boiling-point of oil of turpentine.

Variations in vulcanised or converted waste are sometimes also occasioned by its having been originally formed of two or more of the varieties of caoutchouc, and we have found that those samples that contain any considerable proportion of the softer kinds, or those that in their ordinary state are more readily liquefied by heat, produce solutions that are proportionably more fluid.

Although we have stated that boiling oil of turpentine answers the purpose very well, yet we would observe that a lower temperature may be employed in some cases, such as last mentioned; for instance, when, if the vulcanising or converting has been carried only to a moderate extent, a temperature as low as 250° may suffice, and probably in some cases still lower. The degree of heat employed does not, however, require any nice adjustment; a range from 250° to 312° will not prove injurious in any case, and a little experience will enable the workmen to regulate it according to circumstances.

These vulcanised solutions may be diluted, if desired, with oil of turpentine or coal naphtha; they then become a good and elastic medium for mixing and combining colours.

Instead of operating upon the cuttings or waste of vulcanised or converted caoutchouc, we sometimes take a solution of unvulcanised or unconverted caoutchouc, and mix sulphur with it in the proportion of from 8 to 12 parts

of sulphur to 100 parts of dry caoutchouc, and then submit
the mixture to a temperature of about 300°, or from that to
the boiling point of oil of turpentine, for a period varying
from 15 to 30 minutes, and we thereby obtain a somewhat
similar result; or the caoutchouc, the solvent, and the
sulphur, may be heated in the same manner without
previously dissolving the caoutchouc: but we prefer employing the vulcanised or converted waste, the appro-
priating of which to a useful purpose being the chief
object we have in view.

When the proportion of dry material predominates in
any of these vulcanised solutions so as to form a thick and
plastic mass, they generally become gelatinous, ropy, and
intractable.   We have found that this condition may be
considerably ameliorated or entirely obviated by grinding
or levigation, employing any of the well-know means used
for grinding colours.

We apply these vulcanised solutions to a great variety
of purposes, by coating therewith leather, cloth, and other
fabrics, and thereby rendering them air and waterproof;
and we manufacture from those materials so prepared
various articles, such as pneumatic beds, cushions, and other
similar articles; also cloaks, capes, boots, shoes, and other
articles.   When it is wished to prevent too great an absorption of the more fluid of these vulcanised solutions into
cloth or other materials, we mix or combine with them a
thicker solution of unvulcanised or unconverted caoutchouc, and first apply to the cloth, or other materials, one
or two coatings of such mixture.

We also saturate with the vulcanised solutions, leather,
cloth, felt, and other substances, and manufacture articles
from materials which have been so saturated.   We also use
the vulcanised solutions or cements for uniting together
vulcanised or converted caoutchouc, and fabrics and other
articles.   We also coat with them wood, metal, paper,
plaster casts, and other articles and substances requiring

to be protected from the effects of air and wet. We also employ these vulcanised solutions for taking impressions, by pouring or pressing them into moulds, preferring to use moulds of glass.

We have before mentioned that we mix colouring matters with these vulcanised solutions, and we employ such coloured vulcanised solutions to coat the surfaces of articles and for colouring and printing leather, calico, and other fabrics, and for walls and decorations.

For some purposes we mix with these vulcanised solutions boiled drying oils, black japan, and oil and spirit varnishes. For inferior purposes, or in order to harden or impart other qualities to these vulcanised solutions, we mix or compound with them gums, resins, pitch, asphalte, metallic oxides, earths, wood and cork dust, and fibrous and other substances.

These vulcanised solutions may be applied in any of the modes practised in respect of ordinary paints, varnishes, and solutions of caoutchouc. As they dry somewhat slowly, we prefer exposing them to a temperature of about 180°.

Our improvements in the moulds employed in the manufacture of articles from the vulcanised solutions consist in making such moulds of materials capable of being dissolved or melted, at temperatures so low as not to be injurious to the manufacture; such as the fusible metal known as Darcey's alloy, which melts in boiling water, or compounds of gum, glue, &c. The object in employing these moulds being to facilitate their removal after the vulcanised solutions are dry, by melting or dissolving such moulds as are undercut, or such as have delicate or intricate designs. — In witness, &c.

THOMAS HANCOCK,
REUBEN PHILLIPS.

Enrolled June 30, 1848.

INDEX TO SPECIFICATIONS.

280 INDEX TO SPECIFICATIONS.

Colour to sheets, 269. 277.
Colouring substances, 214, 215. 217. 226. 233. 235. 256, 257. 259. 261. 264, 265. 277.
Combed fibres, 200. 204, 205.
Compound, 248, 249, 250, 251.
Copper, 202.
Cord, 261.
Cord lace and fringe, 255.
Cork dust, 252. 255. 277.
Corn and flour sacks, 204. 214.
Corrugated, or shirred cushions, beds, &c., 218, 219, 220, 221, 222, 223.
Cotton, 196, 197. 203, 204. 206. 208, 209. 212. 215, 216. 232. 251. 271.
Cotton cloth, 205. 250.
Crape, 255. 261.
Cushions, 216. 254. 259. 276.
Cushions, bed, &c., extending, contractile, or corrugated, 218, 219, 220, 221, 222, 223.
Cutting machine, 257.

Darcy's alloy, 277.
Dark colours, 239.
Different degrees of elasticity, 262. 271.
Distillation produces caoutchoucine, 257.
Diving dress, 259.
Dollah, 257. 274.
Double texture, 244.
Dough or putty, 237. 241. 245. 249.
Drab colour, 238.
Draperies, 259.
Drawers, 190.
Dress, 211. 255.

Earths, 277.
Embossing, 233. 250, 251, 252. 255. 257. 259. 261. 266.
Emery, 202. 233.
Engraving, 238. 248. 250. 263.
Engraved rollers, 250.
Epaulettes, 255.

Fancy tubing, 268.
Fastenings for shoes, &c., 261.
Felt, 215. 245. 276.
Fibrous compound, 212, 213. 215. 251.
Fibrous compound, to protect, 259.
Fibres and filaments, 196, 197, 198, 199. 204, 205. 211, 212. 252. 255. 264. 277.

Figures, 211. 213. 216. 232, 233, 234. 249, 250. 252, 253.
Films, or thin sheets, 229, 230, 231.
Fishing-boots, 216. 259.
Flake white, 233.
Flax, 196, 197. 200, 201, 202, 203, 204. 208. 252.
Floor covers, 266.
Forms, 248, 249. 262, 263.
Frames, 216. 262.
Fringe, 261.
Fuller's earth, 238.

Gaiter straps, 190. 216. 260, 261.
Garters, 190. 272.
Glass moulds, 254. 276.
Gloves, 189. 216. 260.
Glue, 211. 213. 232. 268. 277.
Glue size, 206. 212. 220. 233.
Goloshes, 260, 261.
Graining, 232. 238.
Grease, 240. 243.
Gritty matters, 264.
Gum arabic, 212. 232. 277.
Gun or pistol stocks, 263.
Gutta percha, 256, 257, 258, 259, 260. 262, 263, 264, 265. 271. 273, 274.
Gutta percha blocks, 265.
Gutta percha furniture, 265.
Gutta percha machinery straps, 265.
Gutta percha, softness obviated, 258.
Gutta percha sheets, 264, 265.
Gutta percha sheets, to roll evenly, 265.
Gutta percha, to prevent adhering to rollers, 265.
Gutta percha, to purify, 266.
Gutta percha, to dissolve, 273.
Gutta tuban, 257. 274.

Hair, 196, 197. 204, 205. 212.
Handles of whips, 263.
Handles of umbrellas, swords, &c., 263.
Hardening, to expedite, 270.
Harness, 196.
Hard converting, 270.
Hats, 212, 213. 259.
Heat, 193. 205. 219. 225, 226. 228, 229. 243. 245, 246. 249, 250. 265, 266.
Heating unvalcanised solutions, 273.
Hemp, 196. 200. 203, 204, 205, 206. 208. 252.
Hermetically to seal, 266.

THE END.

London :

Printed by Spottiswoode & Co.,
New-street-Square.

Printed in the United States
By Bookmasters